STUDENT
SOLUTIONS MANUAL

ELEMENTARY
LINEAR ALGEBRA

A MATRIX APPROACH

SPENCE • INSEL • FRIEDBERG

Illinois State University

Prentice Hall, Upper Saddle River, NJ 07458

Executive Editor: George Lobell
Editorial Assistant: Gale A. Epps
Special Projects Manager: Barbara A. Murray
Production Editor: Wendy A. Rivers
Supplement Cover Manager: Paul Gourhan
Supplement Cover Designer: PM Workshop Inc.
Manufacturing Buyer: Alan Fischer

Printed in the United States of America

10 9 8 7 6 5 4 3 2 1

ISBN 0-13-025751-6

Prentice-Hall International (UK) Limited, London
Prentice-Hall of Australia Pty. Limited, Sydney
Prentice-Hall Canada, Inc., Toronto
Prentice-Hall Hispanoamericana, S.A., Mexico
Prentice-Hall of India Private Limited, New Delhi
Pearson Education Asia Pte. Ltd., Singapore
Prentice-Hall of Japan, Inc., Tokyo
Editora Prentice-Hall do Brazil, Ltda., Rio de Janeiro

Contents

Chapter 1

Matrices, Vectors, and Systems of Linear Equations

1.1 MATRICES AND VECTORS

1. **(a)** True
   (b) True
   (c) True
   (d) False, a scalar multiple of the zero matrix is the zero matrix.
   (e) False, the transpose of an $m \times n$ matrix is an $n \times m$ matrix.
   (f) True
   (g) False, the rows of B are 1×4 vectors.

5. We have

$$4A - 2B = 4\begin{bmatrix} 2 & -1 & 5 \\ 3 & 4 & 1 \end{bmatrix} - 2\begin{bmatrix} 1 & 0 & -2 \\ 2 & 3 & 4 \end{bmatrix}$$

$$= \begin{bmatrix} 8 & -4 & 20 \\ 12 & 16 & 4 \end{bmatrix} + \begin{bmatrix} -2 & 0 & 4 \\ -4 & -6 & -8 \end{bmatrix}$$

$$= \begin{bmatrix} 6 & -4 & 24 \\ 8 & 10 & -4 \end{bmatrix}.$$

9. -2

13. $\begin{bmatrix} 2 \\ 2e \end{bmatrix}$

17. Let $\mathbf{v}$ be the vector given by the arrow in Figure 1.6. Because the arrow has length 150, we have

$$v_1 = 300 \sin 30° = 150$$
$$v_2 = 300 \cos 30° = 150\sqrt{3}.$$

So $\mathbf{v} = \begin{bmatrix} 150 \\ 150\sqrt{3} \end{bmatrix}$ mph. In $\mathcal{R}^3$, we use the fact that the speed in the z-direction is 10 mph. So the velocity vector of the plane in $\mathcal{R}^3$ is

$$\begin{bmatrix} 150 \\ 150\sqrt{3} \\ 10 \end{bmatrix} \text{ mph.}$$

21. Suppose that A and B are $m \times n$ matrices.

  (a) Clearly the jth column of $A + B$ and $\mathbf{a}_j + \mathbf{b}_j$ are $m \times 1$ vectors. Now the ith component of the jth column of $A+B$ is the (i,j)-entry of $A+B$, which is $a_{ij}+b_{ij}$. By definition, the ith components of $\mathbf{a}_j$ and $\mathbf{b}_j$ are a_{ij} and b_{ij}, respectively. So the ith component of $\mathbf{a}_j + \mathbf{b}_j$ is also $a_{ij} + b_{ij}$. The result follows immediately.

  (b) The jth column of cA and $c\mathbf{a}_j$ are $m \times 1$ vectors. The ith component of the jth column of cA is the (i,j)-entry of cA, which is ca_{ij}. The ith component of $c\mathbf{a}_j$ is also ca_{ij}. So the result follows immediately.

25. The matrices $1A$ and A are clearly the same size, so we need only show they have equal corresponding entries. The (i,j)-entry of $1A$ is $1a_{ij} = a_{ij}$, which is the (i,j)-entry of A.

29. The matrices $(sA)^T$ and sA^T are $n \times m$ matrices; so we need only show they have equal corresponding entries. The (i,j)-entry of $(sA)^T$ is the (j,i)-entry of sA, which is sa_{ji}. The (i,j)-entry of sA^T is the product of s and the (i,j)-entry of A^T, which is sa_{ji}.

33. If $i \neq j$, then the (i,j)-entry of B^T is $b_{ji} = 0$ because $j \neq i$. So B^T is a diagonal matrix.

37. The (i,j)-entry of O is zero, whereas the (i,j)-entry of O^T is the (j,i)-entry of O, which is also zero. So $O = O^T$.

41. No. Consider $\begin{bmatrix} 2 & 5 & 6 \\ 5 & 7 & 8 \\ 6 & 8 & 4 \end{bmatrix}$ and $\begin{bmatrix} 2 & 6 \\ 5 & 8 \end{bmatrix}$.

45. Let $A_1 = \frac{1}{2}(A + A^T)$ and $A_2 = \frac{1}{2}(A - A^T)$. It is easy to show that $A = A_1 + A_2$. By Exercises 39, $A + A^T$ is symmetric. Apply Exercise 38 to this matrix to obtain that $A_1 = \frac{1}{2}(A + A^T)$ is also symmetric. By (k), (j), and (l) of Theorem 1.1, we have

$$A_2^T = \frac{1}{2}(A - A^T)^T = \frac{1}{2}(A^T - (A^T)^T)$$

2

$$= \frac{1}{2}(A^T - A) = -\frac{1}{2}(A - A^T) = -A_2.$$

1.2 LINEAR COMBINATIONS, MATRIX-VECTOR PRODUCTS, AND SPECIAL MATRICES

1. (a) True

   (b) False. Consider the linear combination $3\begin{bmatrix}1\\1\end{bmatrix} + (-3)\begin{bmatrix}1\\1\end{bmatrix} = \begin{bmatrix}0\\0\end{bmatrix}$. If the coefficients were positive, the sum could not equal the zero vector.

   (c) True

   (d) True

   (e) True

   (f) False, the matrix-vector product is a 2×1 vector.

   (g) False, the matrix-vector product is a linear combination of the *columns* of the matrix.

5. We have

$$\begin{bmatrix}2 & -1 & 3\\1 & 0 & -1\\0 & 2 & 4\end{bmatrix}\begin{bmatrix}2\\1\\2\end{bmatrix} = \begin{bmatrix}(2)(2) + (-1)(1) + (3)(2)\\(1)(2) + (0)(1) + (-1)(2)\\(0)(2) + (2)(1) + (4)(2)\end{bmatrix} = \begin{bmatrix}9\\0\\10\end{bmatrix}.$$

9. We have

$$\begin{bmatrix}3 & 0\\2 & 1\end{bmatrix}^T\begin{bmatrix}4\\5\end{bmatrix} = \begin{bmatrix}3 & 2\\0 & 1\end{bmatrix}\begin{bmatrix}4\\5\end{bmatrix} = \begin{bmatrix}(3)(4) + (2)(5)\\(0)(4) + (1)(5)\end{bmatrix} = \begin{bmatrix}22\\5\end{bmatrix}.$$

13. We have

$$\left(\begin{bmatrix}3 & 0\\-2 & 4\end{bmatrix}^T + \begin{bmatrix}1 & 2\\3 & -3\end{bmatrix}^T\right)\begin{bmatrix}4\\5\end{bmatrix} = \left(\begin{bmatrix}3 & -2\\0 & 4\end{bmatrix} + \begin{bmatrix}1 & 3\\2 & -3\end{bmatrix}\right)\begin{bmatrix}4\\5\end{bmatrix}$$

$$= \begin{bmatrix}4 & 1\\2 & 1\end{bmatrix}\begin{bmatrix}4\\5\end{bmatrix} = \begin{bmatrix}(4)(4) + (1)(5)\\(2)(4) + (1)(5)\end{bmatrix} = \begin{bmatrix}21\\13\end{bmatrix}.$$

17. We have

$$A_{60°} = \begin{bmatrix}\cos 60° & -\sin 60°\\\sin 60° & \cos 60°\end{bmatrix} = \begin{bmatrix}\frac{1}{2} & -\frac{\sqrt{3}}{2}\\\frac{\sqrt{3}}{2} & \frac{1}{2}\end{bmatrix} = \frac{1}{2}\begin{bmatrix}1 & -\sqrt{3}\\\sqrt{3} & 1\end{bmatrix},$$

and hence

$$A_{60°}\mathbf{u} = \frac{1}{2}\begin{bmatrix}1 & -\sqrt{3}\\\sqrt{3} & 1\end{bmatrix}\begin{bmatrix}3\\1\end{bmatrix} = \frac{1}{2}\begin{bmatrix}3 - \sqrt{3}\\3\sqrt{3} + 1\end{bmatrix}.$$

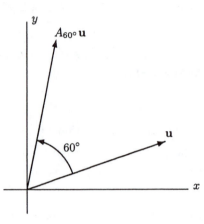

Figure for Exercise 17

21. Let $\mathbf{v} = \begin{bmatrix} a \\ b \end{bmatrix}$. Then

$$A_\theta(A_\beta \mathbf{v}) = \begin{bmatrix} \cos\theta & -\sin\theta \\ \sin\theta & \cos\theta \end{bmatrix} \left(\begin{bmatrix} \cos\beta & -\sin\beta \\ \sin\beta & \cos\beta \end{bmatrix} \begin{bmatrix} a \\ b \end{bmatrix} \right)$$

$$= \begin{bmatrix} \cos\theta & -\sin\theta \\ \sin\theta & \cos\theta \end{bmatrix} \begin{bmatrix} a\cos\beta - b\sin\beta \\ a\sin\beta + b\cos\beta \end{bmatrix}$$

$$= \begin{bmatrix} a\cos\theta\cos\beta - b\cos\theta\sin\beta - a\sin\theta\sin\beta - b\sin\theta\cos\beta \\ a\sin\theta\cos\beta - b\sin\theta\sin\beta + a\cos\theta\sin\beta + b\cos\theta\cos\beta \end{bmatrix}$$

$$= \begin{bmatrix} a\cos(\theta+\beta) - b\sin(\theta+\beta) \\ a\sin(\theta+\beta) + b\cos(\theta+\beta) \end{bmatrix}$$

$$= A_{\theta+\beta}\mathbf{v}.$$

25. the vector $\mathbf{u}$ is not a linear combination of the vectors of $\mathcal{S}$. If $\mathbf{u}$ were a linear combination of $\begin{bmatrix} 4 \\ -4 \end{bmatrix}$, then there would be a scalar c such that

$$\begin{bmatrix} -1 \\ 1 \end{bmatrix} = c \begin{bmatrix} 4 \\ 4 \end{bmatrix} = \begin{bmatrix} 4c \\ 4c \end{bmatrix}.$$

But then $1 = 4c$ and $-1 = 4c$. This is impossible.

29. We seek scalars x_1 and x_2 such that

$$\begin{bmatrix} -1 \\ 11 \end{bmatrix} = x_1 \begin{bmatrix} 1 \\ 3 \end{bmatrix} + x_2 \begin{bmatrix} 2 \\ -1 \end{bmatrix} = \begin{bmatrix} x_1 \\ 3x_1 \end{bmatrix} + \begin{bmatrix} 2x_2 \\ -x_2 \end{bmatrix} = \begin{bmatrix} x_1 + 2x_2 \\ 3x_1 - x_2 \end{bmatrix};$$

that is, we seek a solution to the following system of linear equations:

$$x_1 + 2x_2 = -1$$
$$3x_1 - x_2 = 11.$$

Because these equations represent nonparallel lines in the plane, there is exactly one solution, namely, $x_1 = 3$ and $x_2 = -2$. So

$$\begin{bmatrix} -1 \\ 11 \end{bmatrix} = 3\begin{bmatrix} 1 \\ 3 \end{bmatrix} - 2\begin{bmatrix} 2 \\ -1 \end{bmatrix}.$$

33. Let $\mathbf{p} = \begin{bmatrix} 400 \\ 300 \end{bmatrix}$ and $A = \begin{bmatrix} .85 & .03 \\ .15 & .97 \end{bmatrix}$.

   (a) We compute

$$A\mathbf{p} = \begin{bmatrix} .85 & .03 \\ .15 & .97 \end{bmatrix}\begin{bmatrix} 400 \\ 300 \end{bmatrix}$$

$$= \begin{bmatrix} (.85)(400) + (.03)(300) \\ (.15)(400) + (.97)(300) \end{bmatrix} = \begin{bmatrix} 349 \\ 351 \end{bmatrix}.$$

Thus there are 349,000 in the city and 351,000 in the suburbs.

   (b) We compute the result using (a).

$$A(A\mathbf{p}) = \begin{bmatrix} .85 & .03 \\ .15 & .97 \end{bmatrix}\begin{bmatrix} 349 \\ 351 \end{bmatrix}$$

$$= \begin{bmatrix} (.85)(349) + (.03)(351) \\ (.15)(349) + (.97)(351) \end{bmatrix} = \begin{bmatrix} 307.18 \\ 392.82 \end{bmatrix}.$$

Thus there are 307,180 in the city and 392,820 in the suburbs.

37. The reflection of $\mathbf{u}$ about the x-axis is the vector $\begin{bmatrix} a \\ -b \end{bmatrix}$. To obtain this vector, let $B = \begin{bmatrix} 1 & 0 \\ 0 & -1 \end{bmatrix}$. Then

$$B\mathbf{u} = \begin{bmatrix} 1 & 0 \\ 0 & -1 \end{bmatrix}\begin{bmatrix} a \\ b \end{bmatrix} = \begin{bmatrix} a \\ -b \end{bmatrix}.$$

41. Let $\mathbf{v} = \begin{bmatrix} a \\ 0 \end{bmatrix}$. Then $A\mathbf{v} = \begin{bmatrix} 1 & 0 \\ 0 & 0 \end{bmatrix}\begin{bmatrix} a \\ 0 \end{bmatrix} = \begin{bmatrix} a \\ 0 \end{bmatrix} = \mathbf{v}$.

45. Write $\mathbf{v} = a_1\mathbf{u}_1 + a_2\mathbf{u}_2$ and $\mathbf{w} = b_1\mathbf{u}_1 + b_2\mathbf{u}_2$. Then a typical linear combination of $\mathbf{v}$ and $\mathbf{w}$ is

$$c\mathbf{v} + d\mathbf{w} = c(a_1\mathbf{u}_1 + a_2\mathbf{u}_2) + d(b_1\mathbf{u}_1 + b_2\mathbf{u}_2)$$

$$= (ca_1 + db_1)\mathbf{u}_1 + (ca_2 + db_2)\mathbf{u}_2,$$

which is also a linear combination of $\mathbf{u}_1$ and $\mathbf{u}_2$.

49. We have

$$
\begin{aligned}
(A + B)\mathbf{u} &= u_1(\mathbf{a}_1 + \mathbf{b}_1) + u_2(\mathbf{a}_2 + \mathbf{b}_2) + \cdots + u_n(\mathbf{a}_n + \mathbf{b}_n) \\
&= u_1\mathbf{a}_1 + u_1\mathbf{b}_1 + u_2\mathbf{a}_2 + u_2\mathbf{b}_2 + \cdots + u_n\mathbf{a}_n + u_n\mathbf{b}_n \\
&= (u_1\mathbf{a}_1 + u_2\mathbf{a}_2 + \cdots + u_n\mathbf{a}_n) + (u_1\mathbf{b}_1 + u_2\mathbf{b}_2 + \cdots + u_n\mathbf{b}_n) \\
&= A\mathbf{u} + B\mathbf{u}.
\end{aligned}
$$

53. The jth column of I_n is $\mathbf{e}_j$. So

$$I_n\mathbf{v} = v_1\mathbf{e}_1 + v_2\mathbf{e}_2 + \cdots + v_n\mathbf{e}_n = \mathbf{v}.$$

1.3   SYSTEMS OF LINEAR EQUATIONS

1. **(a)** False, the system $0x_1 + 0x_2 = 1$ has no solutions.
   (b) False, see the boxed result on page 24.
   (c) True
   (d) False, the matrix $\begin{bmatrix} 2 & 0 \\ 0 & 0 \end{bmatrix}$ is in row echelon form.
   (e) True
   (f) True
   (g) False, the matrices $\begin{bmatrix} 2 & 0 \\ 0 & 0 \end{bmatrix}$ and $\begin{bmatrix} 3 & 0 \\ 0 & 0 \end{bmatrix}$ are both row echelon forms for $\begin{bmatrix} 1 & 0 \\ 0 & 0 \end{bmatrix}$.
   (h) True
   (i) True
   (j) False, the system

$$
\begin{aligned}
0x_1 + 0x_2 &= 1 \\
0x_1 + 0x_2 &= 0
\end{aligned}
$$

   is inconsistent, but has augmented matrix $\begin{bmatrix} 0 & 0 & 1 \\ 0 & 0 & 0 \end{bmatrix}$.
   (k) True

5. **(a)** $\begin{bmatrix} 1 & 2 \\ -1 & 3 \\ -3 & 4 \end{bmatrix}$ **(b)** $\begin{bmatrix} 1 & 2 & 3 \\ -1 & 3 & 2 \\ -3 & 4 & 1 \end{bmatrix}$

9. $\begin{bmatrix} 1 & -1 & 0 & 2 & -3 \\ 0 & 4 & 3 & 3 & -5 \\ 0 & 2 & -4 & 4 & 2 \end{bmatrix}$

13.
$$\begin{bmatrix} -2 & 4 & 0 \\ -1 & 1 & -1 \\ 2 & -4 & 6 \\ -3 & 2 & 1 \end{bmatrix}$$

17.
$$\begin{bmatrix} 1 & -2 & 0 \\ 2 & -4 & 6 \\ -1 & 1 & -1 \\ -3 & 2 & 1 \end{bmatrix}$$

21. No, because the left side of the second equation yields $1(2) - 2(1) = 0 \neq -3$. Alternatively,

$$\begin{bmatrix} 1 & -4 & 0 & 3 \\ 0 & 0 & 1 & -2 \end{bmatrix} \begin{bmatrix} 3 \\ 0 \\ 2 \\ 1 \end{bmatrix} = \begin{bmatrix} 6 \\ 0 \end{bmatrix} \neq \begin{bmatrix} 6 \\ -3 \end{bmatrix}.$$

25. The system of linear equations is consistent because the augmented matrix contains no row where the only nonzero entry lies in the last column. The corresponding system of linear equations is

$$\begin{aligned} x_1 - 2x_2 &= 6 \\ 0x_1 + 0x_2 &= 0. \end{aligned}$$

The general solution is

$$\begin{aligned} x_1 &= 6 + 2x_2 \\ x_2 & \quad \text{free.} \end{aligned}$$

29. The system of linear equations is consistent because the augmented matrix contains no row where the only nonzero entry lies in the last column. The corresponding system of linear equations is

$$\begin{aligned} x_1 - 2x_2 \qquad &= 4 \\ x_3 &= 3 \\ 0x_1 + 0x_2 + 0x_3 &= 0. \end{aligned}$$

The general solution is

$$\begin{aligned} x_1 &= 4 + 2x_2 \\ x_2 & \quad \text{free} \\ x_3 &= 3. \end{aligned}$$

33. The system of linear equations is consistent because the augmented matrix contains no row where the only nonzero entry lies in the last column. The corresponding system of linear equations is

$$
\begin{aligned}
x_2 \quad\quad &= -3 \\
x_3 \quad &= -4 \\
x_4 &= 5.
\end{aligned}
$$

The general solution is

$$
\begin{aligned}
x_1 \quad &\text{free} \\
x_2 &= -3 \\
x_3 &= -4 \\
x_4 &= 5 \, .
\end{aligned}
$$

The solution in vector form is

$$
\begin{bmatrix} x_1 \\ x_2 \\ x_3 \\ x_4 \end{bmatrix} = x_1 \begin{bmatrix} 1 \\ 0 \\ 0 \\ 0 \end{bmatrix} + \begin{bmatrix} 0 \\ -3 \\ -4 \\ 5 \end{bmatrix}.
$$

37. The system of linear equations is not consistent because the second row of the augmented matrix has its only nonzero entry in the last column.

41. Clearly, if $[R \ \ c]$ is in reduced row echelon form, then so is R. If we apply the same row operations to A that were applied to $[A \ \ b]$ to produce $[R \ \ c]$, we obtain the matrix R. So R is the reduced row echelon form of A.

45. The ranks of the possible reduced row echelon forms are 0, 1, and 2. Considering each of these ranks, we see that there are 7 possible reduced row echelon forms: $\begin{bmatrix} 0 & 0 & 0 \\ 0 & 0 & 0 \end{bmatrix}$,

$\begin{bmatrix} 1 & * & * \\ 0 & 0 & 0 \end{bmatrix}$, $\begin{bmatrix} 0 & 1 & * \\ 0 & 0 & 0 \end{bmatrix}$, $\begin{bmatrix} 0 & 0 & 1 \\ 0 & 0 & 0 \end{bmatrix}$, $\begin{bmatrix} 1 & 0 & * \\ 0 & 1 & * \end{bmatrix}$, $\begin{bmatrix} 1 & * & 0 \\ 0 & 0 & 1 \end{bmatrix}$, and $\begin{bmatrix} 0 & 1 & 0 \\ 0 & 0 & 1 \end{bmatrix}$.

49. Multiplying the second equation by c produces a system whose augmented matrix is obtained from the augmented matrix of the original system by the elementary row operation of multiplying the second row by c. From the statement on page 27, the two systems are equivalent.

1.4 GAUSSIAN ELIMINATION

1. (a) True

   (b) False. For example, the matrix $\begin{bmatrix} 0 & 1 \\ 2 & 0 \end{bmatrix}$ can be reduced to I_2 by interchanging its rows and then multiplying the first row by $\frac{1}{2}$, or by multiplying the second row by $\frac{1}{2}$ and then interchanging the rows.

   (c) True

   (d) True

   (e) True

   (f) True

   (g) False, from the boxed statement on page 43, we conclude that $\operatorname{rank} A + \operatorname{nullity} A$ equals the number of columns of A. For a 5×8 matrix, we have $3 + 2 \neq 8$.

   (h) False, we need only repeat one equation to produce an equivalent system with a different number of equations.

   (i) True

   (j) True

   (k) True

   (l) False, the augmented matrix $\begin{bmatrix} 1 & 0 & 2 \\ 0 & 1 & 3 \\ 0 & 0 & 0 \end{bmatrix}$ has a zero row, but the corresponding system has the unique solution $x_1 = 2$, $x_2 = 3$.

   (m) False, the augmented matrix $\begin{bmatrix} 0 & 0 & 1 \\ 0 & 0 & 0 \end{bmatrix}$ contains a zero row, but the system is not consistent.

5. The augmented matrix of the given system is

$$\begin{bmatrix} 1 & -2 & -6 \\ -2 & 3 & 7 \end{bmatrix}.$$

Apply the Gaussian elimination algorithm to this augmented matrix to obtain a matrix in reduced row echelon form:

$$\begin{bmatrix} 1 & -2 & -6 \\ -2 & 3 & 7 \end{bmatrix} \longrightarrow \begin{bmatrix} 1 & -2 & -6 \\ 0 & -1 & -5 \end{bmatrix} \longrightarrow \begin{bmatrix} 1 & -2 & -6 \\ 0 & 1 & 5 \end{bmatrix}$$

$$\longrightarrow \begin{bmatrix} 1 & 0 & 4 \\ 0 & 1 & 5 \end{bmatrix}.$$

This matrix corresponds to the system $\begin{array}{c} x_1 = 4 \\ x_2 = 5 \end{array}$, which yields the solution.

9

9. The augmented matrix of this system is

$$\begin{bmatrix} 1 & -2 & -1 & -3 \\ 2 & -4 & 2 & 2 \end{bmatrix}.$$

Apply the Gaussian elimination algorithm to this augmented matrix to obtain a matrix in reduced row echelon form:

$$\begin{bmatrix} 1 & -2 & -1 & -3 \\ 2 & -4 & 2 & 2 \end{bmatrix} \longrightarrow \begin{bmatrix} 1 & -2 & -1 & -3 \\ 0 & 0 & 4 & 8 \end{bmatrix} \longrightarrow \begin{bmatrix} 1 & -2 & -1 & -3 \\ 0 & 0 & 1 & 2 \end{bmatrix}$$

$$\longrightarrow \begin{bmatrix} 1 & -2 & 0 & -1 \\ 0 & 0 & 1 & 2 \end{bmatrix}.$$

This matrix corresponds to the system

$$\begin{aligned} x_1 - 2x_2 \quad &= -1 \\ x_3 &= 2. \end{aligned}$$

Its general solution is

$$\begin{aligned} x_1 &= -1 + 2x_2 \\ x_2 &\quad \text{free} \\ x_3 &= 2. \end{aligned}$$

13. The augmented matrix of this system is

$$\begin{bmatrix} 1 & 3 & 1 & 1 & -1 \\ -2 & -6 & -1 & 0 & 5 \\ 1 & 3 & 2 & 3 & 2 \end{bmatrix}.$$

Apply the Gaussian elimination algorithm to this augmented matrix to obtain a matrix in reduced row echelon form:

$$\begin{bmatrix} 1 & 3 & 1 & 1 & -1 \\ -2 & -6 & -1 & 0 & 5 \\ 1 & 3 & 2 & 3 & 2 \end{bmatrix} \longrightarrow \begin{bmatrix} 1 & 3 & 1 & 1 & -1 \\ 0 & 0 & 1 & 2 & 3 \\ 0 & 0 & 1 & 2 & 3 \end{bmatrix}$$

$$\longrightarrow \begin{bmatrix} 1 & 3 & 1 & 1 & -1 \\ 0 & 0 & 1 & 2 & 3 \\ 0 & 0 & 0 & 0 & 0 \end{bmatrix} \longrightarrow \begin{bmatrix} 1 & 3 & 0 & -1 & -4 \\ 0 & 0 & 1 & 2 & 3 \\ 0 & 0 & 0 & 0 & 0 \end{bmatrix}.$$

This matrix corresponds to the system

$$\begin{aligned} x_1 + 3x_2 \quad - \quad x_4 &= -4 \\ x_3 + 2x_4 &= 3. \end{aligned}$$

Its general solution is

$$
\begin{aligned}
x_1 &= -4 - 3x_2 + x_4 \\
x_2 &\quad \text{free} \\
x_3 &= 3 - 2x_4 \\
x_4 &\quad \text{free.}
\end{aligned}
$$

17. The augmented matrix of this system is

$$
\begin{bmatrix}
1 & 0 & -1 & -2 & -8 & -3 \\
-2 & 0 & 1 & 2 & 9 & 5 \\
3 & 0 & -2 & -3 & -15 & -9
\end{bmatrix}.
$$

Apply the Gaussian elimination algorithm to this augmented matrix to obtain a matrix in reduced row echelon form:

$$
\begin{bmatrix}
1 & 0 & -1 & -2 & -8 & -3 \\
-2 & 0 & 1 & 2 & 9 & 5 \\
3 & 0 & -2 & -3 & -15 & -9
\end{bmatrix}
\longrightarrow
\begin{bmatrix}
1 & 0 & -1 & -2 & -8 & -3 \\
0 & 0 & -1 & -2 & -7 & -1 \\
0 & 0 & 1 & 3 & 9 & 0
\end{bmatrix}
\longrightarrow
$$

$$
\begin{bmatrix}
1 & 0 & -1 & -2 & -8 & -3 \\
0 & 0 & -1 & -2 & -7 & -1 \\
0 & 0 & 0 & 1 & 2 & -1
\end{bmatrix}
\longrightarrow
\begin{bmatrix}
1 & 0 & -1 & 0 & -4 & -5 \\
0 & 0 & -1 & 0 & -3 & -3 \\
0 & 0 & 0 & 1 & 2 & -1
\end{bmatrix}
\longrightarrow
$$

$$
\begin{bmatrix}
1 & 0 & -1 & 0 & -4 & -5 \\
0 & 0 & 1 & 0 & 3 & 3 \\
0 & 0 & 0 & 1 & 2 & -1
\end{bmatrix}
\longrightarrow
\begin{bmatrix}
1 & 0 & 0 & 0 & -1 & -2 \\
0 & 0 & 1 & 0 & 3 & 3 \\
0 & 0 & 0 & 1 & 2 & -1
\end{bmatrix}.
$$

This matrix corresponds to the system

$$
\begin{aligned}
x_1 & & & -x_5 &= -2 \\
& x_3 & & + 3x_5 &= 3 \\
& & x_4 & + 2x_5 &= -1.
\end{aligned}
$$

Its general solution is

$$
\begin{aligned}
x_1 &= -2 + x_5 \\
x_2 &\quad \text{free} \\
x_3 &= 3 - 3x_5 \\
x_4 &= -1 - 2x_5 \\
x_5 &\quad \text{free.}
\end{aligned}
$$

21. The augmented matrix of the system is

$$
\begin{bmatrix}
1 & -2 & 0 \\
4 & -8 & r
\end{bmatrix}.
$$

We apply one elementary row operation and obtain

$$\begin{bmatrix} 1 & -2 & 0 \\ 0 & 0 & r \end{bmatrix}.$$

So the system is inconsistent if $r \neq 0$.

25. The augmented matrix of the system is

$$\begin{bmatrix} 1 & r & 5 \\ 3 & 6 & s \end{bmatrix}.$$

We apply one elementary row operation and obtain

$$\begin{bmatrix} 1 & r & 5 \\ 0 & 6-3r & s-15 \end{bmatrix}.$$

(a) For the system to be inconsistent, we need $6 - 3r = 0$ and $s - 15 \neq 0$. So $r = 2$ and $s \neq 15$.

(b) For the system to have a unique solution, we need $6 - 3r \neq 0$, that is, $r \neq 2$.

(c) For the system to have infinitely many solutions, there must be a free variable. Thus we need $6 - 3r = 0$ and $s - 15 = 0$. So $r = 2$ and $s = 15$.

29. To find the rank and nullity of the given matrix, we first find its reduced row echelon form R:

$$\begin{bmatrix} 1 & -1 & -1 & 0 \\ 2 & -1 & -2 & 1 \\ 1 & -2 & -2 & 2 \\ -4 & 2 & 3 & 1 \\ 1 & -1 & -2 & 3 \end{bmatrix} \longrightarrow \begin{bmatrix} 1 & -1 & -1 & 0 \\ 0 & 1 & 0 & 1 \\ 0 & -1 & -1 & 2 \\ 0 & -2 & -1 & 1 \\ 0 & 0 & -1 & 3 \end{bmatrix} \longrightarrow \begin{bmatrix} 1 & -1 & -1 & 0 \\ 0 & 1 & 0 & 1 \\ 0 & 0 & -1 & 3 \\ 0 & 0 & -1 & 3 \\ 0 & 0 & -1 & 3 \end{bmatrix} \longrightarrow$$

$$\begin{bmatrix} 1 & -1 & -1 & 0 \\ 0 & 1 & 0 & 1 \\ 0 & 0 & -1 & 3 \\ 0 & 0 & 0 & 0 \\ 0 & 0 & 0 & 0 \end{bmatrix} \longrightarrow \begin{bmatrix} 1 & -1 & -1 & 0 \\ 0 & 1 & 0 & 1 \\ 0 & 0 & 1 & -3 \\ 0 & 0 & 0 & 0 \\ 0 & 0 & 0 & 0 \end{bmatrix} \longrightarrow \begin{bmatrix} 1 & -1 & 0 & -3 \\ 0 & 1 & 0 & 1 \\ 0 & 0 & 1 & -3 \\ 0 & 0 & 0 & 0 \\ 0 & 0 & 0 & 0 \end{bmatrix} \longrightarrow$$

$$\begin{bmatrix} 1 & 0 & 0 & -2 \\ 0 & 1 & 0 & 1 \\ 0 & 0 & 1 & -3 \\ 0 & 0 & 0 & 0 \\ 0 & 0 & 0 & 0 \end{bmatrix} = R.$$

The rank of the given matrix equals the number of nonzero rows in R, that is, 3. The nullity of the given matrix equals $n - \operatorname{rank} R = 4 - 3 = 1$.

33. Let x_1, x_2, and x_3 be the number of days that mines 1, 2, and 3, respectively, must operate to supply the desired amounts.

(a) The requirements may be written with the matrix equation

$$\begin{bmatrix} 1 & 1 & 2 \\ 1 & 2 & 2 \\ 2 & 1 & 0 \end{bmatrix} \begin{bmatrix} x_1 \\ x_2 \\ x_3 \end{bmatrix} = \begin{bmatrix} 80 \\ 100 \\ 40 \end{bmatrix}.$$

The reduced row echelon form of the augmented matrix is

$$\begin{bmatrix} 1 & 0 & 0 & 10 \\ 0 & 1 & 0 & 20 \\ 0 & 0 & 1 & 25 \end{bmatrix};$$

so $x_1 = 10$, $x_2 = 20$, $x_3 = 25$.

(b) A similar system of equations yields the reduced row echelon form

$$\begin{bmatrix} 1 & 0 & 0 & 10 \\ 0 & 1 & 0 & 60 \\ 0 & 0 & 1 & -15 \end{bmatrix}.$$

Because $x_3 = -15$ is impossible for this problem, the answer is no.

37. We need $f(-1) = 14$, $f(1) = 4$, and $f(3) = 10$. These conditions produce the system

$$\begin{aligned} a - b + c &= 14 \\ a + b + c &= 4 \\ 9a + 3b + c &= 10, \end{aligned}$$

which has the solution $a = 2$, $b = -5$, $c = 7$. So $f(x) = 2x^2 - 5x + 7$.

41. Column j is e_3. Each pivot column of the reduced row echelon form of A has exactly one nonzero entry, which is 1, and hence it is a standard vector. Also, because of the definition of the reduced row echelon form, the pivot columns in order are $e_1, e_2, \dots$. Hence, the third pivot column must be e_3.

45. The largest possible rank is 4. The reduced row echelon form is a 4×7 matrix and hence has at most 4 nonzero rows. So the rank must be less than or equal to 4. On the other hand, the 4×7 matrix whose first four columns are the distinct standard vectors has rank 4.

49. The largest possible rank is the minimum of m and n. If $m \leq n$, the solution is similar to that of Exercise 45. Suppose that A is an $m \times n$ matrix with $n \leq m$. Every nonzero row of the reduced row echelon form R of A must contain a pivot position, and the column that contains that pivot position is a standard vector. Because R has n columns, it contains at most n columns that are standard vectors, and hence rank $A \leq n$. If every column of R is a distinct standard vector, then rank $A = n$.

53. There are either no solutions or infinitely many solutions. Let the underdetermined system be $A\mathbf{x} = \mathbf{b}$, and let R be the reduced row echelon form of A. Each nonzero row of R corresponds to a basic variable. Since there are fewer equations than variables, there must be free variables. Therefore the system is either inconsistent or has infinitely many solutions.

57. Let A be an $m \times n$ matrix with rank m, $\mathbf{b}$ be a vector in $\mathcal{R}^m$, and $[R \ \mathbf{c}]$ be the reduced row echelon form of $[A \ \mathbf{b}]$. Then R is the reduced row echelon form of A, and hence R has no zero rows. It follows that $[R \ \mathbf{c}]$ has no row in which the only nonzero entry lies in the last column. So $A\mathbf{x} = \mathbf{b}$ is consistent by Theorem 1.4.

61. We have $A(\mathbf{u} + \mathbf{v}) = A\mathbf{u} + A\mathbf{v} = \mathbf{b} + \mathbf{0} = \mathbf{b}$; so $\mathbf{u} + \mathbf{v}$ is a solution to $A\mathbf{x} = \mathbf{b}$.

65.
$$\begin{aligned}
x_1 &= 2.32 + 0.32x_5 \\
x_2 &= -6.44 + 0.56x_5 \\
x_3 &= 0.72 - 0.28x_5 \\
x_4 &= 5.92 + 0.92x_5 \\
x_5 &\quad \text{free}
\end{aligned}$$

69. The reduced row echelon form of the given matrix is (approximately)

$$\begin{bmatrix}
1.0000 & 0 & 0 & 0 & 0 \\
0 & 1.0000 & 0 & 0 & 1.0599 \\
0 & 0 & 1.000 & 0 & 0.8441 \\
0 & 0 & 0 & 1.000 & 0.4925 \\
0 & 0 & 0 & 0 & 0
\end{bmatrix}.$$

The rank equals the number of nonzero rows, 4, and the nullity is found by subtracting the rank from the number of columns, and hence equals $5 - 4 = 1$.

1.5   APPLICATIONS OF SYSTEMS OF LINEAR EQUATIONS

1. (a) False, the net production vector is $\mathbf{x} - C\mathbf{x}$. The vector $C\mathbf{x}$ is the total output of the economy that is consumed during the production process.
   (b) True
   (c) True
   (d) False, see Kirchoff's voltage law.
   (e) True

5. The third column of C gives the amounts from the various sectors required to produce one unit of services. The smallest entry in this column, .06, corresponds to the input from the service sector, and hence services is least dependent on the service sector.

9. Let

$$\mathbf{x} = \begin{bmatrix} 30 \\ 40 \\ 30 \\ 20 \end{bmatrix}.$$

Then

$$C\mathbf{x} = \begin{bmatrix} .12 & .11 & .15 & .18 \\ .20 & .08 & .24 & .07 \\ .18 & .16 & .06 & .22 \\ .09 & .07 & .12 & .05 \end{bmatrix} \begin{bmatrix} 30 \\ 40 \\ 30 \\ 20 \end{bmatrix} = \begin{bmatrix} 16.1 \\ 17.8 \\ 18.0 \\ 10.1 \end{bmatrix}.$$

Therefore the total value of inputs from each sector consumed during the production process are \$16.1 million of agriculture, \$17.8 million of manufacturing, \$18 million of services, and \$10.1 million of entertainment.

13. (a) The gross production vector is $\mathbf{x} = \begin{bmatrix} 40 \\ 30 \\ 35 \end{bmatrix}$. If C is the input-output matrix, then the net production vector is

$$\mathbf{x} - C\mathbf{x} = \begin{bmatrix} 40 \\ 30 \\ 35 \end{bmatrix} - \begin{bmatrix} .2 & .20 & .3 \\ .4 & .30 & .1 \\ .2 & .25 & .3 \end{bmatrix} \begin{bmatrix} 40 \\ 30 \\ 35 \end{bmatrix}$$

$$= \begin{bmatrix} 15.5 \\ 1.5 \\ 9.0 \end{bmatrix}.$$

So the net productions are \$15.5 million of transportation, \$1.5 million of food, and \$9 million of oil.

(b) Let

$$\mathbf{d} = \begin{bmatrix} 32 \\ 48 \\ 24 \end{bmatrix},$$

the net production vector, and let $\mathbf{x}$ denote the gross production vector. Then $\mathbf{x}$ is a solution to the system of linear equations (written in matrix form) $(I_3 - C)\mathbf{x} = \mathbf{d}$. Since

$$I_3 - C = \begin{bmatrix} 1 & 0 & 0 \\ 0 & 1 & 0 \\ 0 & 0 & 1 \end{bmatrix} - \begin{bmatrix} .20 & .20 & .30 \\ .40 & .30 & .10 \\ .20 & .25 & .30 \end{bmatrix} = \begin{bmatrix} .80 & -.20 & -.30 \\ -.40 & .70 & -.10 \\ -.20 & -.25 & .70 \end{bmatrix},$$

the augmented matrix of this system is

$$\begin{bmatrix} .80 & -.20 & -.30 & 32 \\ -.40 & .70 & -.10 & 48 \\ -.20 & -.25 & .70 & 24 \end{bmatrix}.$$

The reduced row echelon form of the augmented matrix is

$$\begin{bmatrix} 1 & 0 & 0 & 128 \\ 0 & 1 & 0 & 160 \\ 0 & 0 & 1 & 128 \end{bmatrix},$$

and hence the gross productions required are \$128 million of transportation, \$160 million of food, and \$128 million of oil.

17. The input-output matrix for this economy is

$$C = \begin{bmatrix} .10 & .10 & .15 \\ .20 & .40 & .10 \\ .20 & .20 & .30 \end{bmatrix}.$$

(a) Let

$$\mathbf{x} = \begin{bmatrix} 70 \\ 50 \\ 60 \end{bmatrix}.$$

Then the net production vector is given by

$$\mathbf{x} - C\mathbf{x} = \begin{bmatrix} 70 \\ 50 \\ 60 \end{bmatrix} - \begin{bmatrix} .10 & .10 & .15 \\ .20 & .40 & .10 \\ .20 & .20 & .30 \end{bmatrix} \begin{bmatrix} 70 \\ 50 \\ 60 \end{bmatrix} = \begin{bmatrix} 49 \\ 10 \\ 18 \end{bmatrix}.$$

Therefore the net productions are \$49 million of finance, \$10 million of goods, and \$18 million of services.

(b) Let

$$\mathbf{d} = \begin{bmatrix} 40 \\ 50 \\ 30 \end{bmatrix},$$

the net production vector, and let $\mathbf{x}$ denote the gross production vector. Then $\mathbf{x}$ is the solution to the matrix equation $(I_3 - C)\mathbf{x} = \mathbf{d}$. Since

$$I_3 - C = \begin{bmatrix} 1 & 0 & 0 \\ 0 & 1 & 0 \\ 0 & 0 & 1 \end{bmatrix} - \begin{bmatrix} .10 & .10 & .15 \\ .20 & .40 & .10 \\ .20 & .20 & .30 \end{bmatrix} = \begin{bmatrix} .90 & -.10 & -.15 \\ -.20 & .60 & -.10 \\ -.20 & -.20 & .70 \end{bmatrix},$$

the augmented matrix of this system is

$$\begin{bmatrix} .90 & -.10 & -.15 & 40 \\ -.20 & .60 & -.10 & 50 \\ -.20 & -.20 & .70 & 30 \end{bmatrix}.$$

The reduced row echelon form of the augmented matrix is

$$\begin{bmatrix} 1 & 0 & 0 & 75 \\ 0 & 1 & 0 & 125 \\ 0 & 0 & 1 & 100 \end{bmatrix},$$

and hence the gross productions are \$75 million of finance, \$125 million of goods, and \$100 million of services.

(c) We proceed as in (b), except that in this case

$$\mathbf{d} = \begin{bmatrix} 40 \\ 36 \\ 44 \end{bmatrix}.$$

The augmented matrix of the system of linear equations $(I_3 - C)\mathbf{x} = \mathbf{d}$ is

$$\begin{bmatrix} .90 & -.10 & -.15 & 40 \\ -.20 & .60 & -.10 & 36 \\ -.20 & -.20 & .70 & 44 \end{bmatrix},$$

which has the reduced row echelon form

$$\begin{bmatrix} 1 & 0 & 0 & 75 \\ 0 & 1 & 0 & 104 \\ 0 & 0 & 1 & 114 \end{bmatrix}.$$

Therefore the gross productions are \$75 million of finance, \$104 million of goods, and \$114 million of services.

21.

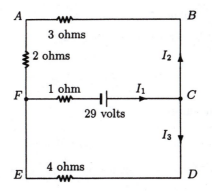

Figure for Exercise 21

Applying Kirchoff's voltage law to the closed path $FCBAF$ in the network above, we obtain the equation

$$3I_2 + 2I_2 + 1I_1 = 29.$$

Similarly, from the closed path $FCDEF$, we obtain

$$1I_1 + 4I_3 = 29.$$

At the junction C, Kirchoff's current law yields the equation

$$I_1 = I_2 + I_3.$$

Thus the currents I_1, I_2, and I_3 satisfy the system

$$\begin{array}{rcr} I_1 + 5I_2 & = & 29 \\ I_1 \phantom{{}+5I_2} + 4I_3 & = & 29 \\ I_1 - I_2 - I_3 & = & 0. \end{array}$$

Since the reduced row echelon form of the augmented matrix of this system is

$$\begin{bmatrix} 1 & 0 & 0 & 9 \\ 0 & 1 & 0 & 4 \\ 0 & 0 & 1 & 5 \end{bmatrix},$$

this system has the unique solution $I_1 = 9$, $I_2 = 4$, $I_3 = 5$.

25.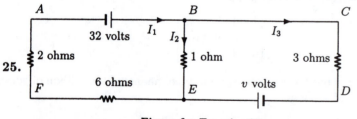

Figure for Exercise 25

Applying Kirchoff's voltage law to the closed path $ABEFA$ in the network above, we obtain the equation

$$1I_2 + 6I_1 + 2I_1 = 32.$$

Similarly, from the closed path $BCDEB$, we obtain

$$1(-I_2) + 3I_3 = v.$$

At junction B, Kirchoff's current law yields the equation

$$I_1 = I_2 + I_3.$$

Therefore the currents I_1, I_2, and I_3 satisfy the system

$$
\begin{array}{rcl}
8I_1 + I_2 & = & 32 \\
-I_2 + 3I_3 & = & v \\
I_1 - I_2 - I_3 & = & 0.
\end{array}
$$

The reduced row echelon form of the augmented matrix of this system is

$$
\begin{bmatrix}
1 & 0 & 0 & \frac{v+128}{35} \\
0 & 1 & 0 & \frac{-8v+96}{35} \\
0 & 0 & 1 & \frac{9v+32}{35}
\end{bmatrix}.
$$

Thus if $I_2 = 0$, we must have $-8v + 96 = 0$, that is, $v = 12$.

1.6   THE SPAN OF A SET OF VECTORS

1. **(a)** True
   (b) True
   (c) True
   (d) False, by Theorem 1.5(c), we need rank $A = m$.
   (e) True
   (f) True
   (g) True
   (h) False, the sets $S_1 = \{e_1\}$ and $S_2 = \{2e_1\}$ have the same spans, but are not equal.
   (i) False, the sets $S_1 = \{e_1\}$ and $S_2 = \{e_1, 2e_1\}$ have equal spans, but do not contain the same number of elements.
   (j) False, $S = \{e_1\}$ and $S \cup \{2e_1\}$ have equal spans, but $2e_1$ is not in S.

5. Let

$$
A = \begin{bmatrix} 1 & -1 & 1 \\ 0 & 1 & 1 \\ 1 & 1 & 3 \end{bmatrix}
\quad \text{and} \quad
v = \begin{bmatrix} 0 \\ 5 \\ 2 \end{bmatrix}.
$$

Then v is in the span if and only if the system $Ax = v$ is consistent. The reduced row echelon form of the augmented matrix of this system is

$$
R = \begin{bmatrix} 1 & 0 & 2 & 0 \\ 0 & 1 & 1 & 0 \\ 0 & 0 & 0 & 1 \end{bmatrix}.
$$

Because of the form of the third row of R, the system is inconsistent. Hence v is not in the span.

9. By the boxed result on page 62, $\mathbf{v}$ is in the span of $\mathcal{S}$ if and only if the system $A\mathbf{x} = \mathbf{v}$ is consistent, where

$$A = \begin{bmatrix} 1 & -1 \\ 0 & 3 \\ -1 & 2 \end{bmatrix}.$$

The augmented matrix of $A\mathbf{v} = \mathbf{0}$ is

$$\begin{bmatrix} 1 & -1 & 2 \\ 0 & 3 & r \\ -1 & 2 & -1 \end{bmatrix}.$$

Applying elementary row operations to this matrix, we obtain

$$\begin{bmatrix} 1 & -1 & 2 \\ 0 & 3 & r \\ -1 & 2 & -1 \end{bmatrix} \longrightarrow \begin{bmatrix} 1 & -1 & 2 \\ 0 & 3 & r \\ 0 & 1 & 1 \end{bmatrix} \longrightarrow \begin{bmatrix} 1 & -1 & 2 \\ 0 & 1 & 1 \\ 0 & 3 & r \end{bmatrix} \longrightarrow$$

$$\begin{bmatrix} 1 & -1 & 2 \\ 0 & 1 & 1 \\ 0 & 0 & r-3 \end{bmatrix}.$$

So the system is consistent if and only if $r - 3 = 0$, that is, $r = 3$. Therefore $\mathbf{v}$ is in the span of $\mathcal{S}$ if and only if $r = 3$.

13. No. Let $A = \begin{bmatrix} 1 & -2 \\ -1 & 2 \end{bmatrix}$. The reduced row echelon form of A is $\begin{bmatrix} 1 & -2 \\ 0 & 0 \end{bmatrix}$, which has rank 1. By Theorem 1.5, the set does not span $\mathcal{R}^2$.

17. Yes. Let

$$A = \begin{bmatrix} 1 & -1 & 1 \\ 0 & 1 & 2 \\ -2 & 4 & -2 \end{bmatrix}.$$

The reduced row echelon form of A is

$$\begin{bmatrix} 1 & 0 & 0 \\ 0 & 1 & 0 \\ 0 & 0 & 1 \end{bmatrix},$$

which has rank 3. By Theorem 1.5, the set spans $\mathcal{R}^3$.

21. Yes. The reduced row echelon form of A is $\begin{bmatrix} 1 & 0 \\ 0 & 1 \end{bmatrix}$, which has rank 2. By Theorem 1.5, the system $A\mathbf{x} = \mathbf{b}$ is consistent for every $\mathbf{b}$ in $\mathcal{R}^2$.

25. The reduced row echelon form of A is

$$\begin{bmatrix} 1 & 0 \\ 0 & 1 \\ 0 & 0 \end{bmatrix},$$

which has rank 2. By Theorem 1.5, $A\mathbf{x} = \mathbf{b}$ is inconsistent for at least one $\mathbf{b}$ in $\mathcal{R}^3$.

29. The desired set is $\left\{ \begin{bmatrix} 1 \\ 3 \end{bmatrix}, \begin{bmatrix} 0 \\ 1 \end{bmatrix} \right\}$. If we delete either vector, then the span of S consists of all multiples of the remaining vector. Because neither vector in S is a multiple of the other, neither can be deleted.

33. The desired set is $\left\{ \begin{bmatrix} 1 \\ -2 \\ 1 \end{bmatrix} \right\}$. The last two vectors in S are multiples of the first, and so can be deleted without changing the span of S.

37. (a) 2
(b) There are infinitely many vectors because every choice of the scalars a and b yields a different vector $a\mathbf{u}_1 + b\mathbf{u}_2$ in the span.

41. Let $S_1 = \{\mathbf{u}_1, \mathbf{u}_2, \ldots, \mathbf{u}_k\}$ and $S_2 = \{\mathbf{u}_1, \mathbf{u}_2, \ldots, \mathbf{u}_r\}$, where $r \geq k$, and suppose that S_1 spans $\mathcal{R}^n$. Let $\mathbf{v}$ be in $\mathcal{R}^n$. Then for some scalars $a_1, a_2, \ldots, a_k$ we can write

$$\mathbf{v} = a_1\mathbf{u}_1 + a_2\mathbf{u}_2 + \cdots + a_k\mathbf{u}_k$$
$$= a_1\mathbf{u}_1 + a_2\mathbf{u}_2 + \cdots + a_k\mathbf{u}_k + 0\mathbf{u}_{k+1} + \cdots + 0\mathbf{u}_r.$$

So S_2 also spans $\mathcal{R}^n$.

45. No, let $A = \begin{bmatrix} 1 & 0 \\ 1 & 0 \end{bmatrix}$. Then $R = \begin{bmatrix} 1 & 0 \\ 0 & 0 \end{bmatrix}$. The span of the columns of A equals all multiples of $\begin{bmatrix} 1 \\ 1 \end{bmatrix}$, whereas the span of the columns of R equals all multiples of $\mathbf{e}_1$.

49. Let $\mathbf{u}_1, \mathbf{u}_2, \ldots, \mathbf{u}_m$ be the rows of A. We must prove that the span of $\mathbf{u}_1, \mathbf{u}_2, \ldots, \mathbf{u}_m$ remains unchanged if we perform any of the three types of elementary row operations. For ease of notation, we will consider operations that affect only the first two rows of A.

Case 1 Suppose we multiply the first row of A by $k \neq 0$. Then the rows of B are $k\mathbf{u}_1, \mathbf{u}_2, \ldots, \mathbf{u}_m$. Any vector in the span of the rows of A can be written

$$c_1\mathbf{u}_1 + c_2\mathbf{u}_2 + \cdots + c_m\mathbf{u}_m = \left(\frac{c_1}{k}\right)k\mathbf{u}_1 + c_2\mathbf{u}_2 + \cdots + c_m\mathbf{u}_m,$$

which is in the span of the rows of B. Likewise, any vector in the span of the rows of B can be written

$$c_1(k\mathbf{u}_1) + c_2\mathbf{u}_2 + \cdots + c_m\mathbf{u}_m = (c_1 k)\mathbf{u}_1 + c_2\mathbf{u}_2 + \cdots + c_m\mathbf{u}_m,$$

which is in the span of the rows of A.

Case 2   Suppose we add k times the second row of A to the first row of A. Then the rows of B are $\mathbf{u}_1 + k\mathbf{u}_2, \mathbf{u}_2, \ldots, \mathbf{u}_m$. Any vector spanned by the rows of A can be can be written

$$c_1\mathbf{u}_1 + c_2\mathbf{u}_2 + \cdots + c_m\mathbf{u}_m = c_1(\mathbf{u}_1 + k\mathbf{u}_2) + (c_2 - kc_1)\mathbf{u}_2 + \cdots + c_m\mathbf{u}_m,$$

which is in the span of the rows of B. Likewise, any vector in the span of the rows of B can be written

$$c_1(\mathbf{u}_1 + k\mathbf{u}_2) + c_2\mathbf{u}_2 + \cdots + c_m\mathbf{u}_m = c_1\mathbf{u}_1 + (kc_1 + c_2)\mathbf{u}_2 + \cdots + c_m\mathbf{u}_m,$$

which is in the span of the rows of A.

Case 3   The rows of A are the same as the rows of B (although in a different order), and hence the span of the rows of A equals the span of the rows of B.

53. no

1.7   LINEAR DEPENDENCE AND LINEAR INDEPENDENCE

1. (a) True

(b) False, the columns are linearly independent (see Theorem 1.7). Consider the matrix $\begin{bmatrix} 1 & 0 \\ 0 & 1 \\ 0 & 0 \end{bmatrix}$.

(c) False, the columns are linearly independent (see Theorem 1.7). See the matrix in the solution to (b).

(d) True

(e) True

(f) True

(g) False, consider the equation $I_2\mathbf{x} = \mathbf{0}$.

(h) True

(i) False, let $\mathbf{v} = \mathbf{0}$.

(j) False, consider the set $\left\{ \begin{bmatrix} 1 \\ 0 \end{bmatrix}, \begin{bmatrix} 0 \\ 1 \end{bmatrix}, \begin{bmatrix} 0 \\ 0 \end{bmatrix} \right\}$.

(k) False, consider $n = 3$ and the set $\left\{ \begin{bmatrix} 1 \\ 0 \\ 0 \end{bmatrix}, \begin{bmatrix} 2 \\ 0 \\ 0 \end{bmatrix} \right\}$.

(l) True

(m) True

5. No, clearly the first two vectors are linearly independent because neither is a multiple of the other. The third vector is not a linear combination of the first two because its first component is not zero. So by property 3 on page 74, the set of 3 vectors is linearly independent.

9. The desired set is $\left\{ \begin{bmatrix} -3 \\ 2 \\ 0 \end{bmatrix}, \begin{bmatrix} 1 \\ 6 \\ 0 \end{bmatrix} \right\}$ because the third vector is zero, and neither of the first two are multiples of the other.

13. Yes, let

$$A = \begin{bmatrix} 1 & 1 & 1 \\ 2 & -3 & 2 \\ 0 & 1 & -2 \\ -1 & -2 & 3 \end{bmatrix},$$

which has reduced row echelon form

$$\begin{bmatrix} 1 & 0 & 0 \\ 0 & 1 & 0 \\ 0 & 0 & 1 \\ 0 & 0 & 0 \end{bmatrix}.$$

So rank $A = 3$. By Theorem 1.7, the set is linearly independent.

17. No, let

$$A = \begin{bmatrix} 1 & -1 & -1 & 0 \\ -1 & 0 & -4 & 1 \\ -1 & 1 & 1 & -2 \\ 2 & -1 & 3 & 1 \end{bmatrix},$$

which has reduced row echelon form

$$\begin{bmatrix} 1 & 0 & 4 & 0 \\ 0 & 1 & 5 & 0 \\ 0 & 0 & 0 & 1 \\ 0 & 0 & 0 & 0 \end{bmatrix}.$$

So rank $A = 3$. By Theorem 1.7, the set is linearly dependent.

21. Let A be the matrix whose columns are the vectors in $\mathcal{S}$. By Theorem 1.7, there exists a nonzero solution to $A\mathbf{x} = \mathbf{0}$. The general solution to this system is

$$\begin{aligned} x_1 &= -5x_3 \\ x_2 &= -4x_3 \\ x_3 &\quad \text{free.} \end{aligned}$$

So one solution is $x_1 = -5$, $x_2 = -4$, and $x_3 = 1$. Therefore

$$-5 \begin{bmatrix} 0 \\ 1 \\ 1 \end{bmatrix} - 4 \begin{bmatrix} 1 \\ 0 \\ -1 \end{bmatrix} + \begin{bmatrix} 4 \\ 5 \\ 1 \end{bmatrix} = \mathbf{0}.$$

Hence

$$\begin{bmatrix} 4 \\ 5 \\ 1 \end{bmatrix} = 5 \begin{bmatrix} 0 \\ 1 \\ 1 \end{bmatrix} + 4 \begin{bmatrix} 1 \\ 0 \\ -1 \end{bmatrix}.$$

25. Let

$$A = \begin{bmatrix} -2 & 1 & -1 \\ 0 & 1 & 1 \\ 1 & -3 & r \end{bmatrix},$$

which may be reduced to

$$\begin{bmatrix} 1 & 0 & 1 \\ 0 & 1 & 1 \\ 0 & 0 & r+2 \end{bmatrix}.$$

Then rank $A = 2$ if $r = -2$. By Theorem 1.7, the set is linearly dependent if $r = -2$.

29. The general solution is

$$\begin{aligned} x_1 &= -3x_2 - 2x_4 \\ x_2 &\quad \text{free} \\ x_3 &= \quad\quad\quad 6x_4 \\ x_4 &\quad \text{free.} \end{aligned}$$

So

$$\begin{bmatrix} x_1 \\ x_2 \\ x_3 \\ x_4 \end{bmatrix} = \begin{bmatrix} -3x_2 - 2x_4 \\ x_2 \\ 6x_4 \\ x_4 \end{bmatrix} = x_2 \begin{bmatrix} -3 \\ 1 \\ 0 \\ 0 \end{bmatrix} + x_4 \begin{bmatrix} -2 \\ 0 \\ 6 \\ 1 \end{bmatrix}.$$

33. The general solution is

$$\begin{aligned} x_1 &= -x_4 - 3x_6 \\ x_2 &\quad \text{free} \\ x_3 &= \quad 2x_4 - \quad x_6 \\ x_4 &= \text{free} \\ x_5 &= \quad 0 \\ x_6 &\quad \text{free.} \end{aligned}$$

So

$$\begin{bmatrix} x_1 \\ x_2 \\ x_3 \\ x_4 \\ x_5 \\ x_6 \end{bmatrix} = \begin{bmatrix} -x_4 - 3x_6 \\ x_2 \\ 2x_4 - x_6 \\ x_4 \\ 0 \\ x_6 \end{bmatrix} = x_2 \begin{bmatrix} 0 \\ 1 \\ 0 \\ 0 \\ 0 \\ 0 \end{bmatrix} + x_4 \begin{bmatrix} -1 \\ 0 \\ 2 \\ 1 \\ 0 \\ 0 \end{bmatrix} + x_6 \begin{bmatrix} -3 \\ 0 \\ -1 \\ 0 \\ 0 \\ 1 \end{bmatrix}$$

37. If $\mathcal{S} \cup \{\mathbf{v}\}$ is linearly independent, then $\mathbf{v}$ does not belong to the span of $\mathcal{S}$. We prove the equivalent statement: If $\mathbf{v}$ belongs to the span of $\mathcal{S}$, then $\mathcal{S} \cup \{\mathbf{v}\}$ is linearly dependent. Let $\mathcal{S} = \{\mathbf{u}_1, \mathbf{u}_2, \ldots, \mathbf{u}_k\}$. Because $\mathbf{v}$ is in the span of $\mathcal{S}$, we may write

$$\mathbf{v} = c_1 \mathbf{u}_1 + c_2 \mathbf{u}_2 + \cdots + c_k \mathbf{u}_k$$

for some scalars $c_1, c_2, \ldots, c_k$. So

$$c_1 \mathbf{u}_1 + c_2 \mathbf{u}_2 + \cdots + c_k \mathbf{u}_k + (-1)\mathbf{v} = \mathbf{0}.$$

Because not all the coefficients are zero, we may conclude that $\mathcal{S} \cup \{\mathbf{v}\}$ is linearly dependent.

41. Suppose

$$a_1(c_1 \mathbf{u}_1) + a_2(c_2 \mathbf{u}_2) + \cdots + a_k(c_k \mathbf{u}_k) = \mathbf{0}.$$

Then $0 = a_1 c_1 = a_2 c_2 = \cdots = a_k c_k$. Since each c_i is nonzero, $a_1 = a_2 = \cdots = a_k = 0$.

45. Suppose that $\mathbf{v}$ is in the span of $\mathcal{S}$ and that

$$\mathbf{v} = c_1 \mathbf{u}_1 + c_2 \mathbf{u}_2 + \cdots + c_k \mathbf{u}_k \qquad \text{and} \qquad \mathbf{v} = d_1 \mathbf{u}_1 + d_2 \mathbf{u}_2 + \cdots + d_k \mathbf{u}_k$$

for scalars $c_1, c_2, \ldots, c_k, d_1, d_2, \ldots, d_k$. Subtracting the second equation from the first yields

$$\mathbf{0} = (c_1 - d_1)\mathbf{u}_1 + (c_2 - d_2)\mathbf{u}_2 + \cdots + (c_k - d_k)\mathbf{u}_k.$$

Because $\mathcal{S}$ is linearly independent, we have $c_1 - d_1 = c_2 - d_2 = \cdots = c_k - d_k = 0$. So $c_1 = d_1, c_2 = d_2, \cdots, c_k = d_k$.

49. Let $\mathbf{u}_1, \mathbf{u}_2, \ldots, \mathbf{u}_k$ be the nonzero rows of A, and suppose

$$a_1 \mathbf{u}_1 + a_2 \mathbf{u}_2 + \cdots + a_k \mathbf{u}_k = \mathbf{0}.$$

Each $\mathbf{u}_i$ has a component equal to 1 in a position where the other vectors have zeros. By equating components of both sides of the equation corresponding to those positions, we obtain $a_1 = 0, a_2 = 0, \cdots, a_k = 0$. So $\mathbf{u}_1, \mathbf{u}_2, \ldots, \mathbf{u}_k$ are linearly independent.

53. The set is linearly dependent and $\mathbf{v}_5 = 2\mathbf{v}_1 - \mathbf{v}_3 + \mathbf{v}_4$, where $\mathbf{v}_j$ is the jth vector in the set.

CHAPTER 1 REVIEW

1. (a) False, the columns are 3×1 vectors.
   (b) True

(c) True

(d) True

(e) True

(f) True

(g) True

(h) False, the nonzero entry has to be the last entry.

(i) False, consider the matrix in reduced row echelon form $\begin{bmatrix} 1 & 0 & 2 \\ 0 & 1 & 3 \\ 0 & 0 & 0 \end{bmatrix}$. The associated system has the unique solution $x_1 = 2$, $x_2 = 3$.

(j) True

(k) True

(l) True

(m) False, in $A = \begin{bmatrix} 1 & 2 \end{bmatrix}$, the columns are linearly dependent, but rank $A = 1 = m$.

(n) True

(o) True

(p) False, the subset $\left\{ \begin{bmatrix} 1 \\ 2 \\ 3 \end{bmatrix}, \begin{bmatrix} 2 \\ 4 \\ 6 \end{bmatrix} \right\}$ of $\mathcal{R}^3$ is linearly dependent.

(q) False, consider the example in (p).

5. $\begin{bmatrix} 3 & 2 \\ -2 & 7 \\ 4 & 3 \end{bmatrix}$

9. We have

$$A^T D^T = \begin{bmatrix} 1 & -2 & 0 \\ 3 & 4 & 2 \end{bmatrix} \begin{bmatrix} 1 \\ -1 \\ 2 \end{bmatrix} = \begin{bmatrix} (1)(1) + (-2)(-1) + (0)(2) \\ (3)(1) + (4)(-1) + (2)(2) \end{bmatrix} = \begin{bmatrix} 3 \\ 3 \end{bmatrix}.$$

13. The components are the average values of sales at all stores during January of last year for produce, meats, dairy, and processed foods, respectively.

17. We have

$$A_{-30°} \begin{bmatrix} 2 \\ -1 \end{bmatrix} = \begin{bmatrix} \cos(-30°) & -\sin(-30°) \\ \sin(-30°) & \cos(-30°) \end{bmatrix} \begin{bmatrix} 2 \\ -1 \end{bmatrix}$$

$$= \begin{bmatrix} \frac{\sqrt{3}}{2} & \frac{1}{2} \\ -\frac{1}{2} & \frac{\sqrt{3}}{2} \end{bmatrix} \begin{bmatrix} 2 \\ -1 \end{bmatrix} = \frac{1}{2} \begin{bmatrix} \sqrt{3} & 1 \\ -1 & \sqrt{3} \end{bmatrix} \begin{bmatrix} 2 \\ -1 \end{bmatrix}$$

$$= \frac{1}{2} \begin{bmatrix} (\sqrt{3})(2) + (1)(-1) \\ (-1)(2) + (\sqrt{3})(-1) \end{bmatrix} = \frac{1}{2} \begin{bmatrix} 2\sqrt{3} - 1 \\ -2 - \sqrt{3} \end{bmatrix}.$$

21. Let A be the matrix whose columns are the vectors in $\mathcal{S}$. Then $\mathbf{v}$ is a linear combination of the vectors in $\mathcal{S}$ if and only if the system $A\mathbf{x} = \mathbf{v}$ is consistent. The reduced row echelon form of the augmented matrix of this system is

$$\begin{bmatrix} 1 & 0 & -\frac{1}{2} & 0 \\ 0 & 1 & \frac{1}{2} & 0 \\ 0 & 0 & 0 & 1 \end{bmatrix}.$$

Because the third row has its only nonzero entry in the last column, this system is not consistent. So $\mathbf{v}$ is not a linear combination of the vectors in $\mathcal{S}$.

25. The reduced row echelon form of the augmented matrix of the system is

$$\begin{bmatrix} 1 & 0 & -\frac{1}{3} & 0 \\ 0 & 1 & \frac{5}{3} & 0 \\ 0 & 0 & 0 & 1 \end{bmatrix}.$$

Because the third row has its only nonzero entry in the last column, the system is not consistent.

29. The reduced row echelon form is $\begin{bmatrix} 1 & 2 & -3 & 0 & 1 \end{bmatrix}$, and so the rank is 1. Thus the nullity is $5 - 1 = 4$.

33. Let x_1, x_2, x_3, respectively, be the appropriate numbers of the three packs. We must solve the system

$$\begin{aligned} 10x_1 + 10x_2 + 5x_3 &= 500 \\ 10x_1 + 15x_2 + 10x_3 &= 750 \\ 10x_2 + 5x_3 &= 300, \end{aligned}$$

where the first equation represents the total number of oranges from the three packs, the second equation represents the total number of grapefruit from the three packs, and the third equation represents the total number of apples from the three packs. We obtain the solution $x_1 = 20$, $x_2 = 10$, $x_3 = 40$.

37. Let A be the matrix whose columns are the vectors in the given set. Then by Theorem 1.5, the set spans $\mathcal{R}^3$ if and only if rank $A = 3$. The reduced row echelon form of A is

$$\begin{bmatrix} 1 & 0 & \frac{2}{3} & \frac{1}{3} \\ 0 & 1 & \frac{1}{3} & -\frac{1}{3} \\ 0 & 0 & 0 & 0 \end{bmatrix}.$$

Therefore rank $A = 2$, and so the set does not span $\mathcal{R}^3$.

41. Yes. The reduced row echelon form of A is

$$\begin{bmatrix} 1 & 0 & 0 \\ 0 & 1 & 0 \\ 0 & 0 & 1 \end{bmatrix};$$

so rank $A = 3$. By Theorem 1.5, the equation $A\mathbf{x} = \mathbf{b}$ is consistent for every $\mathbf{b}$ in $\mathcal{R}^n$.

45. Let A be the matrix whose columns are the vectors in the given set. By Theorem 1.7, the set is linearly independent if and only if rank $A = 3$. The reduced row echelon form of A is

$$\begin{bmatrix} 1 & 0 & 0 \\ 0 & 1 & 0 \\ 0 & 0 & 1 \\ 0 & 0 & 0 \end{bmatrix},$$

and hence rank $A = 3$. So the set is linearly independent.

49. Let A be the matrix whose columns are the vectors in $\mathcal{S}$. By Theorem 1.7, there exists a nonzero solution to $A\mathbf{x} = \mathbf{0}$. The general solution to this system is

$$\begin{aligned} x_1 &= -2x_3 \\ x_2 &= -x_3 \\ x_3 & \quad \text{free.} \end{aligned}$$

So one solution is $x_1 = -2$, $x_2 = -1$, $x_3 = 1$. Therefore

$$-2 \begin{bmatrix} 1 \\ 2 \\ 3 \end{bmatrix} - \begin{bmatrix} 1 \\ -1 \\ 2 \end{bmatrix} + \begin{bmatrix} 3 \\ 3 \\ 8 \end{bmatrix} = \mathbf{0}.$$

Thus

$$\begin{bmatrix} 3 \\ 3 \\ 8 \end{bmatrix} = 2 \begin{bmatrix} 1 \\ 2 \\ 3 \end{bmatrix} + \begin{bmatrix} 1 \\ -1 \\ 2 \end{bmatrix}.$$

53. The general solution to the system is

$$\begin{aligned} x_1 &= -3x_3 \\ x_2 &= 2x_3 \\ x_3 & \quad \text{free.} \end{aligned}$$

So

$$\begin{bmatrix} x_1 \\ x_2 \\ x_3 \end{bmatrix} = \begin{bmatrix} -3x_3 \\ 2x_3 \\ x_3 \end{bmatrix} = x_3 \begin{bmatrix} -3 \\ 2 \\ 1 \end{bmatrix}.$$

57. We prove the equivalent result: Suppose that $\mathbf{w}_1$ and $\mathbf{w}_2$ are linear combinations of vectors $\mathbf{v}_1$ and $\mathbf{v}_2$. If $\mathbf{v}_1$ and $\mathbf{v}_2$ are linearly dependent, then $\mathbf{w}_1$ and $\mathbf{w}_2$ are linearly dependent. By assumption, one of $\mathbf{v}_1$ or $\mathbf{v}_2$ is a multiple of the other, say $\mathbf{v}_1 = k\mathbf{v}_2$ for some scalar k. Thus we may write

$$\mathbf{w}_1 = a_1\mathbf{v}_1 + a_2\mathbf{v}_2 = a_1k\mathbf{v}_2 + a_2\mathbf{v}_2 = (a_1k + a_2)\mathbf{v}_2$$
$$\mathbf{w}_2 = b_1\mathbf{v}_1 + b_2\mathbf{v}_2 = b_1k\mathbf{v}_2 + b_2\mathbf{v}_2 = (b_1k + b_2)\mathbf{v}_2$$

for some scalars a_1, a_2, b_1, b_2. Let $c_1 = a_1k + a_2$ and $c_2 = b_1k + b_2$. Then $\mathbf{w}_1 = c_1\mathbf{v}_2$ and $\mathbf{w}_2 = c_2\mathbf{v}_2$. Because $\mathbf{w}_1$ and $\mathbf{w}_2$ are linearly independent, $c_1 \neq 0 \neq c_2$. So

$$\mathbf{w}_1 = c_1\mathbf{v}_2 = c_1\left(\frac{1}{c_2}\right)\mathbf{w}_2 = \frac{c_1}{c_2}\mathbf{w}_2,$$

and hence $\mathbf{w}_1$ and $\mathbf{w}_2$ are linearly dependent.

Chapter 2

Matrices and Linear Transformations

2.1 MATRIX MULTIPLICATION

1. (a) False, the product is not defined unless $n = m$.
   (b) True
   (c) False, if A is a 2×3 matrix and B is a 3×2 matrix, then the product is defined.
   (d) False, $(AB)^T = B^T A^T$.
   (e) True
   (f) True
   (g) False, consider the (undefined) product

   $$\left[\begin{array}{cc|c} 1 & 2 & 3 \\ 4 & 5 & 6 \\ \hline 7 & 8 & 9 \end{array}\right] \left[\begin{array}{cc|c} 1 & 2 & 3 \\ 4 & 5 & 6 \\ \hline 7 & 8 & 9 \end{array}\right].$$

5. Let A be any 3×4 matrix and B be any 2×3 matrix. So BA is defined because for $(2 \times 3)(3 \times 4)$ the inner dimensions are equal. But AB is not defined because for $(3 \times 4)(2 \times 3)$ the inner dimensions are not equal.

9. We have

   $$\mathbf{xz} = \begin{bmatrix} 2 \\ 3 \end{bmatrix} [7 \quad -1] = \begin{bmatrix} (2)(7) & (2)(-1) \\ (3)(7) & (3)(-1) \end{bmatrix} = \begin{bmatrix} 14 & -2 \\ 21 & -3 \end{bmatrix}.$$

13. We have

   $$AB = \begin{bmatrix} 1 & -2 \\ 3 & 4 \end{bmatrix} \begin{bmatrix} 7 & 4 \\ 1 & 2 \end{bmatrix} = \begin{bmatrix} (1)(7) + (-2)(1) & (1)(4) + (-2)(2) \\ (3)(7) + (4)(1) & (3)(4) + (4)(2) \end{bmatrix} = \begin{bmatrix} 5 & 0 \\ 25 & 20 \end{bmatrix}.$$

17. CB^T is undefined because C is a 2×3 matrix, B^T is a 2×2 matrix, and $3 \neq 2$.

21. C^2 is undefined because C is a 2×3 matrix and $3 \neq 2$.

25. By item 3 on page 95, the $(3, 2)$-entry of AB equals

$$\mathbf{a'}_3\mathbf{b}_2 = [-3 \ -2 \ 0]\begin{bmatrix} 0 \\ 1 \\ -2 \end{bmatrix} = (-3)(0) + (-2)(1) + (0)(-2) = -2.$$

29. By item 1 on page 95, the second column of AB is

$$A\mathbf{b}_2 = \begin{bmatrix} 1 & 2 & 3 \\ 2 & -1 & 4 \\ -3 & -2 & 0 \end{bmatrix}\begin{bmatrix} 0 \\ 1 \\ -2 \end{bmatrix} = \begin{bmatrix} -4 \\ -9 \\ -2 \end{bmatrix}.$$

33. By item 4 on page 95, we have

$$AB = \mathbf{a}_1\mathbf{b'}_1 + \mathbf{a}_2\mathbf{b'}_2 + \mathbf{a}_3\mathbf{b'}_3$$

$$= \begin{bmatrix} 1 \\ 2 \\ -3 \end{bmatrix}[-1 \ 0] + \begin{bmatrix} 2 \\ -1 \\ -2 \end{bmatrix}[4 \ 1] + \begin{bmatrix} 3 \\ 4 \\ 0 \end{bmatrix}[3 \ -2]$$

$$= \begin{bmatrix} -1 & 0 \\ -2 & 0 \\ 3 & 0 \end{bmatrix} + \begin{bmatrix} 8 & 2 \\ -4 & -1 \\ -8 & -2 \end{bmatrix} + \begin{bmatrix} 9 & -6 \\ 12 & -8 \\ 0 & 0 \end{bmatrix}.$$

37. We have

$$\begin{bmatrix} 3 & 0 \\ 0 & 3 \\ 2 & 0 \\ 0 & 2 \end{bmatrix}\begin{bmatrix} 1 & 2 \\ 3 & 4 \end{bmatrix} = \begin{bmatrix} \begin{bmatrix} 3 & 0 \\ 0 & 3 \end{bmatrix}\begin{bmatrix} 1 & 2 \\ 3 & 4 \end{bmatrix} \\ \begin{bmatrix} 2 & 0 \\ 0 & 2 \end{bmatrix}\begin{bmatrix} 1 & 2 \\ 3 & 4 \end{bmatrix} \end{bmatrix} = \begin{bmatrix} 3 & 6 \\ 9 & 12 \\ 2 & 4 \\ 6 & 8 \end{bmatrix}$$

41. (a) The number of people living in single unit houses is clearly $.70v_1 + .95v_2$, and the number of people living in multiple unit housing is clearly $.30v_1 + .05v_2$. These results may be expressed as the matrix equation

$$\begin{bmatrix} .70 & .95 \\ .30 & .05 \end{bmatrix}\begin{bmatrix} v_1 \\ v_2 \end{bmatrix} = \begin{bmatrix} u_1 \\ u_2 \end{bmatrix}.$$

So take

$$B = \begin{bmatrix} .70 & .95 \\ .30 & .05 \end{bmatrix}.$$

(b) Because $A\begin{bmatrix} v_1 \\ v_2 \end{bmatrix}$ represents the number of people living in the city and suburbs after one year, it follows from (a) that $BA\begin{bmatrix} v_1 \\ v_2 \end{bmatrix}$ gives the number of single unit and multiple unit dwellers after one year in the city and suburbs.

45. We prove that $C(P + Q) = CP + CQ$. Note that $P + Q$ is an $n \times p$ matrix, and so $C(P + Q)$ is an $m \times p$ matrix. Also CP and CQ are both $m \times p$ matrices; so $CP + CQ$ is an $m \times p$ matrix. Hence the matrices on both sides of the equation have the same size. The jth column of $P + Q$ is $\mathbf{p}_j + \mathbf{q}_j$; so the jth column of $C(P + Q)$ is $C(\mathbf{p}_j + \mathbf{q}_j)$, which equals $C\mathbf{p}_j + C\mathbf{q}_j$ by Theorem 1.2(d). On the other hand, the jth columns of CP and CQ are $C\mathbf{p}_j$ and $C\mathbf{q}_j$, respectively. So the jth column of $CP + CQ$ is $C\mathbf{p}_j + C\mathbf{q}_j$. Thus $C(P + Q)$ and $CP + CQ$ have the same corresponding columns, and hence are equal.

49. As on page 94, we have

$$\mathbf{ab}^T = \begin{bmatrix} a_1\mathbf{b}^T \\ a_2\mathbf{b}^T \\ \vdots \\ a_m\mathbf{b}^T \end{bmatrix};$$

so all rows are multiples of $\mathbf{b}^T$. Because $\mathbf{a} \neq \mathbf{0}$, some component of $\mathbf{a}$, say a_i, does not equal zero. We may add multiples of the ith row to other rows to eventually obtain a matrix whose only nonzero row is the ith row of $\mathbf{ab}^T$. For example, we would add $-\frac{a_1}{a_i}$ times row i to row 1. Now we interchange the ith row with the first row and multiply the first row by $\frac{1}{a_i}$. We conclude that the reduced row echelon form of $\mathbf{ab}^T$ has one nonzero row. Therefore rank $(\mathbf{ab}^T) = 1$.

53. The statement is false. Let $A = \begin{bmatrix} 1 & 0 \\ 0 & 0 \end{bmatrix}$ and $B = \begin{bmatrix} 0 & 0 \\ 1 & 0 \end{bmatrix}$. Then $AB = O$, but

$$BA = \begin{bmatrix} 0 & 0 \\ 1 & 0 \end{bmatrix} \neq O.$$

57. Using (b) and (g) of Theorem 2.2, we have

$$(ABC)^T = ((AB)C)^T = C^T(AB)^T = C^T(B^T A^T) = C^T B^T A^T.$$

61. (c) $A^k = \begin{bmatrix} B^k & * \\ 0 & D^k \end{bmatrix}$, where * represents some 2×2 matrix.

(d) We have

$$A^2 = AA = \begin{bmatrix} B & C \\ O & D \end{bmatrix}\begin{bmatrix} B & C \\ O & D \end{bmatrix} = \begin{bmatrix} BB + CO & BC + CD \\ OB + DO & OC + DD \end{bmatrix} = \begin{bmatrix} B^2 & * \\ O & D^2 \end{bmatrix}.$$

So

$$A^3 = A^2 A = \begin{bmatrix} B^2 & * \\ O & D^2 \end{bmatrix}\begin{bmatrix} B & C \\ O & D \end{bmatrix} = \begin{bmatrix} B^2 B + O & * \\ OB + D^2 O & OC + D^2 D \end{bmatrix} = \begin{bmatrix} B^3 & * \\ O & D^3 \end{bmatrix}.$$

2.2   APPLICATIONS OF MATRIX MULTIPLICATION

1. (a) False, the population may be decreasing.
    (b) False, the population may continue to grow without bound.
    (c) True
    (d) True
    (e) True
    (f) False, $\mathbf{z} = BA\mathbf{x}$.
    (g) True
    (h) False, $A = \begin{bmatrix} 0 & 1 \\ 0 & 0 \end{bmatrix}$ is a nonsymmetric adjacency matrix.

5. (a) Using the notation on page 101, we have $p_1 = q$ and $p_2 = .5$. Also $b_1 = 0$ because females under age 1 do not give birth. Likewise, $b_2 = 2$ and $b_3 = 1$. So the Leslie matrix is

$$A = \begin{bmatrix} b_1 & b_2 & b_3 \\ p_1 & 0 & 0 \\ 0 & p_2 & 0 \end{bmatrix} = \begin{bmatrix} 0 & 2 & 1 \\ q & 0 & 0 \\ 0 & .5 & 0 \end{bmatrix}.$$

    (b) Let

$$A = \begin{bmatrix} 0 & 2 & 1 \\ .8 & 0 & 0 \\ 0 & .5 & 0 \end{bmatrix} \quad \text{and} \quad \mathbf{x}_0 = \begin{bmatrix} 300 \\ 1180 \\ 130 \end{bmatrix}.$$

The population in 50 years is given by

$$A^{50}\mathbf{x}_0 \approx 10^9 \begin{bmatrix} 9.28 \\ 5.40 \\ 1.96 \end{bmatrix}.$$

So the population appears to grow without bound.

    (c) With $q = .2$, we have

$$A = \begin{bmatrix} 0 & 2 & 1 \\ .2 & 0 & 0 \\ 0 & .5 & 0 \end{bmatrix}.$$

As in (b), we compute

$$A^{50}\mathbf{x}_0 \approx 10^{-3} \begin{bmatrix} .3697 \\ 1.009 \\ .0689 \end{bmatrix}$$

and conclude that the population appears to approach zero.

33

(d) $q = .4$. The stable distribution is $\begin{bmatrix} 400 \\ 160 \\ 80 \end{bmatrix}$.

(e) For $q = .4$ and $\mathbf{x}_0 = \begin{bmatrix} 210 \\ 240 \\ 180 \end{bmatrix}$, we compute $A^5\mathbf{x}_0$, $A^{10}\mathbf{x}_0$, and $A^{30}\mathbf{x}_0$ obtaining

$$\begin{bmatrix} 513.60 \\ 144.96 \\ 114.00 \end{bmatrix}, \qquad \begin{bmatrix} 437.36 \\ 189.99 \\ 85.17 \end{bmatrix}, \qquad \text{and} \qquad \begin{bmatrix} 499.98 \\ 180.01 \\ 89.99 \end{bmatrix}.$$

It appears that the populations approach

$$\begin{bmatrix} 450 \\ 180 \\ 90 \end{bmatrix}.$$

(f) From Theorem 1.7, we need rank $(A - I_3) < 3$. Using elementary row operations, the matrix

$$A - I_3 = \begin{bmatrix} 0 & 2 & 1 \\ q & 0 & 0 \\ 0 & .5 & 0 \end{bmatrix}$$

may be transformed to

$$\begin{bmatrix} 1 & -2 & -1 \\ 0 & 1 & -2 \\ 0 & 0 & -2(1 - 2q) + q \end{bmatrix}.$$

For rank $(A - I_3) < 3$, we need $-2(1 - 2q) + q = 0$, or $q = .4$. This is the value obtained in (d).

(g) The solution of $(A - I_3)\mathbf{x} = \mathbf{0}$ is $x_3 \begin{bmatrix} 5 \\ 2 \\ 1 \end{bmatrix}$. For $x_3 = 90$, we obtain the solution in (e).

9. Let p and q be the amounts of donations and interest received by the foundation, and let n and a be the net income and fund raising costs, respectively. Then

$$n = .7p + .9q$$
$$a = .3p + .1q,$$

and hence

$$\begin{bmatrix} n \\ a \end{bmatrix} = \begin{bmatrix} .7 & .9 \\ .3 & .1 \end{bmatrix} \begin{bmatrix} p \\ q \end{bmatrix}.$$

Next, let r and c be the amounts of net income used for research and clinic maintenance, respectively. Then

$$r = .4n$$
$$c = .6n$$
$$a = \quad a,$$

and hence

$$\begin{bmatrix} r \\ c \\ a \end{bmatrix} = \begin{bmatrix} .4 & 0 \\ .6 & 0 \\ 0 & 1 \end{bmatrix} \begin{bmatrix} n \\ a \end{bmatrix}.$$

Finally, let m and f be the material and personnel costs of the foundation, respectively. Then

$$m = .8r + .5c + .7a$$
$$f = .2r + .5c + .3a,$$

and hence

$$\begin{bmatrix} m \\ f \end{bmatrix} = \begin{bmatrix} .8 & .5 & .7 \\ .2 & .5 & .4 \end{bmatrix} \begin{bmatrix} r \\ c \\ a \end{bmatrix}.$$

Combining these matrix equations, we have

$$\begin{bmatrix} m \\ f \end{bmatrix} = \begin{bmatrix} .8 & .5 & .7 \\ .2 & .5 & .4 \end{bmatrix} \begin{bmatrix} .4 & 0 \\ .6 & 0 \\ 0 & 1 \end{bmatrix} \begin{bmatrix} n \\ a \end{bmatrix} = \begin{bmatrix} .8 & .5 & .7 \\ .2 & .5 & .4 \end{bmatrix} \begin{bmatrix} .4 & 0 \\ .6 & 0 \\ 0 & 1 \end{bmatrix} \begin{bmatrix} .7 & .9 \\ .3 & .1 \end{bmatrix} \begin{bmatrix} p \\ q \end{bmatrix}$$

$$= \begin{bmatrix} 0.644 & 0.628 \\ 0.356 & 0.372 \end{bmatrix} \begin{bmatrix} p \\ q \end{bmatrix}.$$

13. **(a)** We need only check the (i,j)-entries such that $a_{ij} = a_{ji} = 1$. This yields 1 and 2, 1 and 4, 2 and 3, and 3 and 4.

   (b) The (i,j)-entry of A^2 is

   $$a_{i1}a_{1j} + a_{i2}a_{2j} + a_{i3}a_{3j} + a_{i4}a_{4j}.$$

   The kth term of this sum equals 1 if and only person i likes person k and person k likes person j. Otherwise the term is 0. So the (i,j)-entry of A^2 equals the number of people whom person i likes and who like person j.

   (c) We have

   $$B = \begin{bmatrix} 0 & 1 & 0 & 1 \\ 1 & 0 & 1 & 0 \\ 0 & 1 & 0 & 1 \\ 1 & 0 & 1 & 0 \end{bmatrix}.$$

   Clearly B is symmetric because $B = B^T$.

(d) Because $B^3 = B^2 B$, we have that the (i, i)-entry of B^3 consists of a sum of terms of the form $c_{ik}b_{ki}$, where c_{ik} consists of a sum of terms of the form $b_{ij}b_{jk}$. Therefore the (i, i)-entry of B^3 consists of terms of the form $b_{ij}b_{jk}b_{ki}$. The (i, i)-entry of B^3 is positive if and only if some term $b_{ij}b_{jk}b_{ki} > 0$. This occurs if and only if $b_{ij} = 1 = b_{jk} = b_{ki}$, that is, if and only if there are friends k and j who are also friends of person i, so that person i is in a clique.

(e) We have

$$B^3 = \begin{bmatrix} 0 & 4 & 0 & 4 \\ 4 & 0 & 4 & 0 \\ 0 & 4 & 0 & 4 \\ 4 & 0 & 4 & 0 \end{bmatrix} ;$$

so the (i, i)-entry is 0 for every i. Therefore there are no cliques.

17. Using the notation in this section, we let

$$\mathbf{x}_0 = \begin{bmatrix} 100 \\ 200 \\ 300 \\ 8000 \end{bmatrix} \quad \text{and} \quad A = \begin{bmatrix} 1 & 0 & 0 & 0 \\ 1 & 1 & 0 & 0 \\ 0 & 1 & 1 & 0 \\ -1 & -1 & 0 & 1 \end{bmatrix}.$$

(a) To compute the numbers in each class in k generations, we must compute $\mathbf{x}_k = A\mathbf{x}_{k-1}$ for $k = 1, 2, 3$. We put the results in the table below.

k	Sun	Noble	Honored	Stinkard
1	100	300	500	7700
2	100	400	800	7300
3	100	500	1200	6800

(b)

k	Sun	Noble	Honored	Stinkard
9	100	1100	5700	1700
10	100	1200	6800	500
11	100	1300	8000	-800

(c) The tribe will cease to exist because every member is required to marry a Stinkard, and the number of Stinkards decreases to zero.

(d) We must find k such that $s_k + n_k + h_k > st_k$. That is, there are sufficiently many Stinkards for the other classes to marry. From (8), this inequality is equivalent to

$$s_0 + (n_0 + ks_0) + \left(h_0 + kn_0 + \frac{k(k-1)}{2}s_0 \right) > st_0 - kn_0 - \frac{k(k+1)}{2}s_0.$$

If we let $s_0 = 100$, $n_0 = 200$, $k_0 = 300$, and $st_0 = 8000$ and simplify the inequality above, we obtain $k^2 + 5k - 74 > 0$. The smallest value of k that satisfies this inequality is $k = 7$.

2.3 INVERTIBILITY AND ELEMENTARY MATRICES

1. (a) False, the $n \times n$ zero matrix is not invertible (see page 114).
   (b) True
   (c) True
   (d) False, let

   $$A = \begin{bmatrix} 1 & 0 & 0 \\ 0 & 1 & 0 \end{bmatrix} \quad \text{and} \quad B = \begin{bmatrix} 1 & 0 \\ 0 & 1 \\ 0 & 0 \end{bmatrix}.$$

   Then $AB = I_2$, but neither A nor B is square; so neither is invertible.
   (e) True
   (f) True

5. Yes. We need only show that $AB = BA = I_3$.

9. (a) We compute the product

   $$\left(\frac{1}{ad - bc} \begin{bmatrix} d & -b \\ -c & a \end{bmatrix} \right) \begin{bmatrix} a & b \\ c & d \end{bmatrix} = \frac{1}{ad - bc} \begin{bmatrix} ad - bc & db - bd \\ -ac + ac & -bc + ad \end{bmatrix} = I_2.$$

   The product in the reverse order also yields I_2, so by the uniqueness result on page 114, A is invertible and

   $$A^{-1} = \frac{1}{ad - bc} \begin{bmatrix} d & -b \\ -c & a \end{bmatrix}.$$

   (b) Suppose that $A = \begin{bmatrix} a & b \\ c & d \end{bmatrix}$ is invertible. Then for any $\mathbf{v}$ in $\mathcal{R}^2$, $A^{-1}\mathbf{v}$ is a solution to $A\mathbf{x} = \mathbf{v}$ (see page 115). Thus rank $A = 2$ by Theorem 1.5. It follows that no row or column of A can be $\mathbf{0}$, and so $a \neq 0$ or $c \neq 0$.
   If $a \neq 0$, add $-\frac{c}{a}$ times row 1 of A to row 2 to obtain

   $$\begin{bmatrix} a & b \\ c & d \end{bmatrix} \longrightarrow \begin{bmatrix} a & b \\ 0 & d - \frac{bc}{a} \end{bmatrix}.$$

   Since A has rank 2, we must have $d - \frac{bc}{a} \neq 0$. Multiplication by a therefore gives $ad - bc \neq 0$.
   On the other hand, if $a = 0$, then

   $$A = \begin{bmatrix} 0 & b \\ c & d \end{bmatrix}.$$

   Since A has rank 2, we must have $b \neq 0$ and $c \neq 0$. Thus

   $$ad - bc = 0d - bc = -bc \neq 0.$$

   Hence, in either case, $ad - bc \neq 0$.

37

13. The given matrix is an elementary matrix that corresponds to multiplying the second row by 4. So its inverse is the elementary matrix that corresponds to multiplying the second row by $\frac{1}{4}$. It is given below.

$$\begin{bmatrix} 1 & 0 & 0 & 0 \\ 0 & .25 & 0 & 0 \\ 0 & 0 & 1 & 0 \\ 0 & 0 & 0 & 1 \end{bmatrix}$$

17. By a boxed result on page 117, every elementary matrix is invertible. By the boxed result on page 116, the product of invertible matrices is invertible. The result follows immediately.

21. Using Theorem 2.2(b) and Theorem 2.3(b), we have

$$(ABC)^{-1} = [(AB)C]^{-1} = C^{-1}(AB)^{-1} = C^{-1}(B^{-1}A^{-1}) = C^{-1}B^{-1}A^{-1}.$$

25. By Theorem 1.4, we have that $B\mathbf{x} = \mathbf{b}$ has a solution for every $\mathbf{b}$ in $\mathcal{R}^n$. So for every standard vector $\mathbf{e}_i$ there is a vector $\mathbf{u}_i$ that satisfies $B\mathbf{u}_i = \mathbf{e}_i$. Let $C = [\mathbf{u}_1 \ \mathbf{u}_2 \ \cdots \ \mathbf{u}_n]$. Then

$$BC = [B\mathbf{u}_1 \ B\mathbf{u}_2 \ \cdots \ B\mathbf{u}_n] = [\mathbf{e}_1 \ \mathbf{e}_2 \ \cdots \ \mathbf{e}_n] = I_n.$$

29. From the linear correspondence property, it follows that the third column of A equals

$$\mathbf{a}_1 + 2\mathbf{a}_2 = \begin{bmatrix} 3 \\ -1 \end{bmatrix} + 2 \begin{bmatrix} 2 \\ 5 \end{bmatrix} = \begin{bmatrix} 7 \\ 9 \end{bmatrix}.$$

Therefore $A = \begin{bmatrix} 3 & 2 & 7 \\ -1 & 5 & 9 \end{bmatrix}$.

33. Let R be the reduced row echelon form of A. Because $\mathbf{u}$ and $\mathbf{v}$ are linearly independent, $\mathbf{a}_1 = \mathbf{u} \neq \mathbf{0}$, and hence $\mathbf{a}_1$ is a pivot column. Thus $\mathbf{r}_1 = \mathbf{e}_1$. Since $\mathbf{a}_2 = 2\mathbf{u} = 2\mathbf{a}_1$, it follows that $\mathbf{r}_2 = 2\mathbf{r}_1 = 2\mathbf{e}_1$ by the linear correspondence property. Since $\mathbf{u}$ and $\mathbf{v}$ are linearly independent, it is easy to show that $\mathbf{u}$ and $\mathbf{u} + \mathbf{v}$ are linearly independent, and hence $\mathbf{a}_3$ is not a linear combination of $\mathbf{a}_1$ and $\mathbf{a}_2$. Thus $\mathbf{a}_3$ is a pivot column and so $\mathbf{r}_3 = \mathbf{e}_2$. Finally,

$$\mathbf{a}_4 = \mathbf{a}_3 - \mathbf{u} = \mathbf{a}_3 - \mathbf{a}_1,$$

and hence $\mathbf{r}_4 = \mathbf{r}_3 - \mathbf{r}_1$ by the linear correspondence property. Therefore

$$R = \begin{bmatrix} 1 & 2 & 0 & -1 \\ 0 & 0 & 1 & 1 \\ 0 & 0 & 0 & 0 \end{bmatrix}.$$

37. Suppose that rank $R = r$. Then the reduced row echelon form of R^T is the $n \times m$ matrix with (i, i)-entry equal to 1 for $1 \leq i \leq r$ and zeros elsewhere.

To justify this, first observe that the first r columns of R^T are nonzero and that all subsequent columns of R^T are zero vectors. Next observe that for every j with $1 \leq j \leq r$, the first nonzero entry in the jth column is 1, and it is above the first nonzero entry of each subsequent row. Now perform elementary row operations to convert all of the entries below the first nonzero entry in any column to zero. In the resulting matrix, the first r columns of the matrix are distinct standard vectors of $\mathcal{R}^n$. Now perform row interchanges to obtain the desired result.

41. We first prove the result for a 2×2 matrix A where the row operation is adding k times the second row to the first. Let

$$A = \begin{bmatrix} a & b \\ c & d \end{bmatrix} \quad \text{and} \quad E = \begin{bmatrix} 1 & k \\ 0 & 1 \end{bmatrix}.$$

Then

$$EA = \begin{bmatrix} 1 & k \\ 0 & 1 \end{bmatrix} \begin{bmatrix} a & b \\ c & d \end{bmatrix} = \begin{bmatrix} a + kc & b + kd \\ c & d \end{bmatrix}.$$

More generally, let A be an $m \times n$ matrix. Let $\mathbf{e}'_i$ and $\mathbf{a}'_i$ denote the ith rows of I_m and A, respectively. Observe that

$$\mathbf{e}'_i A = \mathbf{a}'_i,$$

and for any $m \times m$ matrix B,

$$BA = \begin{bmatrix} \mathbf{b}'_1 \\ \mathbf{b}'_2 \\ \vdots \\ \mathbf{b}'_m \end{bmatrix} A = \begin{bmatrix} \mathbf{b}'_1 A \\ \mathbf{b}'_2 A \\ \vdots \\ \mathbf{b}'_m A \end{bmatrix},$$

where $\mathbf{b}'_i$ denotes the ith row of B.

Now consider the elementary row operation on A in which k times the ith row of A is added to the jth row of A. Let E be the corresponding elementary matrix. Then

$$E = \begin{bmatrix} \mathbf{e}'_1 \\ \mathbf{e}'_2 \\ \vdots \\ \mathbf{e}'_j + k\mathbf{e}'_i \\ \vdots \\ \mathbf{e}'_m \end{bmatrix}.$$

Therefore

$$EA = \begin{bmatrix} \mathbf{e'}_1 \\ \mathbf{e'}_2 \\ \vdots \\ \mathbf{e'}_j + k\mathbf{e'}_i \\ \vdots \\ \mathbf{e'}_m \end{bmatrix} A = \begin{bmatrix} \mathbf{e'}_1 A \\ \mathbf{e'}_2 A \\ \vdots \\ (\mathbf{e'}_j + k\mathbf{e'}_i)A \\ \vdots \\ \mathbf{e'}_m A \end{bmatrix} = \begin{bmatrix} \mathbf{a'}_1 \\ \mathbf{a'}_2 \\ \vdots \\ \mathbf{a'}_j + k\mathbf{a'}_i \\ \vdots \\ \mathbf{a'}_m \end{bmatrix},$$

which is the desired result. The proofs for other types of elementary row operations
are similar.

45. (a) $A^{-1} = \begin{bmatrix} -7 & 2 & 3 & -2 \\ 5 & -1 & -2 & 1 \\ 1 & 0 & 0 & 1 \\ -3 & 1 & 1 & -1 \end{bmatrix}$

(b) $B^{-1} = \begin{bmatrix} 3 & 2 & -7 & -2 \\ -2 & -1 & 5 & 1 \\ 0 & 0 & 1 & 1 \\ 1 & 1 & -3 & -1 \end{bmatrix}$ and $C^{-1} = \begin{bmatrix} -7 & -2 & 3 & 2 \\ 5 & 1 & -2 & -1 \\ 1 & 1 & 0 & 0 \\ -3 & -1 & 1 & 1 \end{bmatrix}$

(c) B^{-1} can be obtained by interchanging columns 1 and 3 of A^{-1}, and C^{-1} can be
obtained by interchanging columns 2 and 4 of A^{-1}.

(d) B^{-1} can be obtained by interchanging columns i and j of A^{-1}.

(e) Let E be the elementary matrix that corresponds to interchanging rows i and
j. Then $B = EA$. So by Theorem 2.3(b), we have $B^{-1} = (EA)^{-1} = A^{-1}E^{-1}$.
It follows from Exercise 44 that B^{-1} is obtained from A^{-1} by performing the
elementary column operation on A^{-1} that is associated with E^{-1}. But because
E is associated with a row interchange, we have $E^2 = I_n$, and hence $E^{-1} = E$.

49. (a) See (c) for the inverse.

(b)

$$A^{-1} = \begin{bmatrix} 0.2870 & 0.3912 & 0.2400 & -0.0848 \\ -0.2993 & -0.3759 & -0.1471 & 0.2094 \\ -0.2339 & -0.4576 & -0.3095 & 0.2114 \\ -0.1369 & -0.3289 & -0.1287 & 0.0582 \end{bmatrix},$$

$$B^{-1} = \begin{bmatrix} -0.0384 & 0.0042 & 0.0558 & 0.0740 \\ 0.0813 & 0.0483 & -0.0412 & -0.0335 \\ -0.0188 & -0.0112 & -0.0674 & 0.0846 \\ 0.0784 & -0.0615 & 0.0070 & 0.0426 \end{bmatrix}, \quad \text{and}$$

$$C^{-1} = \begin{bmatrix} -0.0556 & -0.1429 & -0.1032 & 0.1746 \\ 0.0444 & -0.1714 & 0.0540 & 0.0317 \\ 0.0028 & -0.1714 & -0.2377 & 0.1984 \\ 0.0306 & 0.2571 & 0.0996 & -0.1032 \end{bmatrix}$$

(c)

$$-A^{-1}BC^{-1} = \begin{bmatrix} 0.0569 & 1.4822 & -0.3997 & -0.6180 \\ 0.0183 & -1.2210 & 0.5182 & 0.3251 \\ -0.0902 & -1.5286 & 0.5630 & 0.5010 \\ -0.0607 & -1.3683 & 0.9012 & 0.6247 \end{bmatrix}$$

(e) If A, B, and C are invertible $n \times n$ matrices, then $\begin{bmatrix} A & B \\ O & C \end{bmatrix}$ is invertible and

$$\begin{bmatrix} A & B \\ O & C \end{bmatrix}^{-1} = \begin{bmatrix} A^{-1} & -A^{-1}BC^{-1} \\ O & C^{-1} \end{bmatrix}.$$

For a proof, use block multiplication to obtain

$$\begin{bmatrix} A & B \\ O & C \end{bmatrix} \begin{bmatrix} A^{-1} & -A^{-1}BC^{-1} \\ O & C^{-1} \end{bmatrix} = \begin{bmatrix} AA^{-1} + BO & -AA^{-1}BC^{-1} + BC^{-1} \\ OA^{-1} + CO & O + CC^{-1} \end{bmatrix}$$

$$= \begin{bmatrix} I_n & O \\ O & I_n \end{bmatrix}.$$

Since we obtain the same result if we reverse the factors, the proof is complete.

2.4   THE INVERSE OF A MATRIX

1. **(a)** False, let

$$A = \begin{bmatrix} 1 & 0 & 0 \\ 0 & 1 & 0 \end{bmatrix} \quad \text{and} \quad B = \begin{bmatrix} 1 & 0 \\ 0 & 1 \\ 0 & 0 \end{bmatrix}.$$

Then $AB = I_2$, but A is not square; so it is not invertible.
   (b) True
   (c) True
   (d) True
   (e) True
   (f) False, let $A = I_2$ and $B = -I_2$. Then $A + B = O$, which is not invertible.
   (g) True

5. The reduced row echelon form of the given matrix is

$$\begin{bmatrix} 1 & 0 & 1 \\ 0 & 1 & 0 \\ 0 & 0 & 0 \end{bmatrix}.$$

Since this matrix is not I_3, Theorem 2.6 implies that the matrix is not invertible.

7. Let A be the given matrix. The reduced row echelon form of A is I_3, so by Theorem 2.6, A is invertible. To find A^{-1}, we find the reduced row echelon form of $[A \ \ I_3]$:

$$\begin{bmatrix} 1 & 1 & 2 & 1 & 0 & 0 \\ 2 & -1 & 1 & 0 & 1 & 0 \\ 2 & 3 & 4 & 0 & 0 & 1 \end{bmatrix} \longrightarrow \begin{bmatrix} 1 & 1 & 2 & 1 & 0 & 0 \\ 0 & -3 & -3 & -2 & 1 & 0 \\ 0 & 1 & 0 & -2 & 0 & 1 \end{bmatrix} \longrightarrow$$

$$\begin{bmatrix} 1 & 1 & 2 & 1 & 0 & 0 \\ 0 & 1 & 0 & -2 & 0 & 1 \\ 0 & -3 & -3 & -2 & 1 & 0 \end{bmatrix} \longrightarrow \begin{bmatrix} 1 & 1 & 2 & 1 & 0 & 0 \\ 0 & 1 & 0 & -2 & 0 & 1 \\ 0 & 0 & -3 & -8 & 1 & 3 \end{bmatrix} \longrightarrow$$

$$\begin{bmatrix} 1 & 1 & 2 & 1 & 0 & 0 \\ 0 & 1 & 0 & -2 & 0 & 1 \\ 0 & 0 & 1 & \frac{8}{3} & -\frac{1}{3} & -1 \end{bmatrix} \longrightarrow \begin{bmatrix} 1 & 1 & 0 & -\frac{13}{3} & \frac{2}{3} & 2 \\ 0 & 1 & 0 & -2 & 0 & 1 \\ 0 & 0 & 1 & \frac{8}{3} & -\frac{1}{3} & -1 \end{bmatrix} \longrightarrow$$

$$\begin{bmatrix} 1 & 0 & 0 & -\frac{7}{3} & \frac{2}{3} & 1 \\ 0 & 1 & 0 & -2 & 0 & 1 \\ 0 & 0 & 1 & \frac{8}{3} & -\frac{1}{3} & -1 \end{bmatrix}$$

So

$$A^{-1} = \frac{1}{3}\begin{bmatrix} -7 & 2 & 3 \\ -6 & 0 & 3 \\ 8 & -1 & -3 \end{bmatrix}.$$

9. The reduced row echelon form of the given matrix is

$$\begin{bmatrix} 1 & 0 & 0 & 1 \\ 0 & 1 & 0 & 1 \\ 0 & 0 & 1 & -1 \\ 0 & 0 & 0 & 0 \end{bmatrix}.$$

Since this matrix is not I_4, Theorem 2.6 implies that the given matrix is not invertible.

13. $A^{-1}B = \begin{bmatrix} -1 & -4 & 7 & -7 \\ 2 & 6 & -6 & 10 \end{bmatrix}$. Using the algorithm on page 130, we form the matrix

$$\begin{bmatrix} 2 & 2 & 2 & 4 & 2 & 6 \\ 2 & 1 & 0 & -2 & 8 & -4 \end{bmatrix}$$

and compute its reduced row echelon form

$$\left[\begin{array}{cc|cccc} 1 & 0 & -1 & 4 & 7 & -7 \\ 0 & 1 & 2 & 6 & -6 & 10 \end{array}\right] = [I_2 \ A^{-1}B].$$

17. $R = \begin{bmatrix} 1 & 0 & -1 \\ 0 & 1 & -3 \end{bmatrix}$ and $P = \begin{bmatrix} -1 & -1 \\ -2 & -1 \end{bmatrix}$. By the discussion on page 128, the reduced row echelon form of $[A \ I_n]$ is $[R \ P]$. In this problem,

$$[R \ P] = \left[\begin{array}{ccc|cc} 1 & 0 & -1 & -1 & -1 \\ 0 & 1 & -3 & -2 & -1 \end{array}\right].$$

21. $R = \begin{bmatrix} 1 & 0 & 0 & 0 \\ 0 & 1 & 0 & 0 \\ 0 & 0 & 1 & 0 \\ 0 & 0 & 0 & 1 \end{bmatrix}$ and $P = \begin{bmatrix} -4 & -15 & -8 & 1 \\ 1 & 4 & 2 & 0 \\ 1 & 3 & 2 & 0 \\ -4 & -13 & -7 & 1 \end{bmatrix}$. By the discussion on page 128, the reduced row echelon form of $[A \ I_n]$ is $[R \ P]$. In this problem,

$$[R \ P] = \left[\begin{array}{cccc|cccc} 1 & 0 & 0 & 0 & -4 & -15 & -8 & 1 \\ 0 & 1 & 0 & 0 & 1 & 4 & 2 & 0 \\ 0 & 0 & 1 & 0 & 1 & 3 & 2 & 0 \\ 0 & 0 & 0 & 1 & -4 & -13 & -7 & 1 \end{array}\right].$$

25. (a) $\begin{bmatrix} -1 & -3 \\ 2 & 5 \end{bmatrix} \begin{bmatrix} x_1 \\ x_2 \end{bmatrix} = \begin{bmatrix} -6 \\ 4 \end{bmatrix}$

(b) We determine the reduced row echelon form of

$$\left[\begin{array}{cc|cc} -1 & -3 & 1 & 0 \\ 2 & 5 & 0 & 1 \end{array}\right]$$

to be

$$\left[\begin{array}{cc|cc} 1 & 0 & 5 & 3 \\ 0 & 1 & -2 & -1 \end{array}\right];$$

so $A^{-1} = \begin{bmatrix} 5 & 3 \\ -2 & -1 \end{bmatrix}$.

(c) By Exercise 23(a), we have

$$\begin{bmatrix} x_1 \\ x_2 \end{bmatrix} = A^{-1}\mathbf{b} = \begin{bmatrix} 5 & 3 \\ -2 & -1 \end{bmatrix} \begin{bmatrix} -6 \\ 4 \end{bmatrix} = \begin{bmatrix} -18 \\ 8 \end{bmatrix}.$$

29. (a) If $k = 1$, then $A = I_n$, and the result is obvious. If $k > 1$, rewrite the equation $A^k = I_n$ as $A(A^{k-1}) = I_n$. Theorem 2.6 shows that A is invertible.

(b) From (a) and the fact that inverses are unique, we have $A^{-1} = A^{k-1}$.

33. No. Let $A = \begin{bmatrix} 2 & 4 \\ 0 & 0 \end{bmatrix}$. Then $R = \begin{bmatrix} 1 & 2 \\ 0 & 0 \end{bmatrix}$. For

$$P = \begin{bmatrix} \frac{1}{2} & 0 \\ 0 & 1 \end{bmatrix} \quad \text{and} \quad Q = \begin{bmatrix} \frac{1}{2} & 0 \\ 0 & 3 \end{bmatrix},$$

we have $PA = R$ and $QA = R$, but $P \neq Q$.

37. Clearly conditions 2 and 3 of the definition of reduced row echelon form on page 28 are satisfied for R if they are satisfied for $[R \ S]$. Let $A = [R \ S]$. To show that condition 1 is satisfied, suppose that $\mathbf{r}_i = \mathbf{0}$ but for some $j > i$, $\mathbf{r}_j \neq \mathbf{0}$. Because A is in reduced row echelon form, $\mathbf{a}_i \neq \mathbf{0}$ since $\mathbf{a}_j \neq \mathbf{0}$. So the leading nonzero entry of $\mathbf{a}_i$ is to the right of the leading nonzero entry of $\mathbf{a}_j$. This contradicts condition 2 for A.

41. In Exercise 15(c) of Section 1.5, we have two sectors, oil and electricity. The input-output matrix is given by

$$C = \begin{bmatrix} .1 & .4 \\ .3 & .2 \end{bmatrix}.$$

As in Example 5, we need to compute 3 times the second column of $(I_2 - C)^{-1}$. Since

$$3(I_2 - C)^{-1} = 3 \begin{bmatrix} \frac{4}{3} & \frac{2}{3} \\ \frac{1}{2} & \frac{3}{2} \end{bmatrix} = \begin{bmatrix} 4.0 & 2.0 \\ 1.5 & 4.5 \end{bmatrix},$$

the amount required is \$2 million of electricity and \$4.5 million of oil.

45. Suppose the net production of sector i must be increased by k units, where $k > 0$. The gross production vector is given by

$$(I_n - C)^{-1}\mathbf{d} + k\mathbf{p}_i,$$

where C is the input-output matrix, $\mathbf{d}$ is the original demand vector, and $\mathbf{p}_i$ is the ith column of $(I_n - C)^{-1}$. If all the entries of $\mathbf{p}_i$ are positive, then the gross production of every sector of the economy must be increased.

49. (a) We are given that $I_n = P^{-1}AP$ for some invertible matrix P. It follows that

$$A = PI_nP^{-1} = PP^{-1} = I_n.$$

(b) We are given that $O = P^{-1}AP$ for some invertible matrix P. So $POP^{-1} = A$, and $A = O$.

(c) A is similar to $B = cI_n$ if there is an invertible matrix P such that $cI_n = P^{-1}AP$. So

$$A = P(cI_n)P^{-1} = cPI_nP^{-1} = cPP^{-1} = cI_n.$$

Thus $A = B$.

53. The reduced row echelon form of A is I_4.

2.5 THE LU DECOMPOSITION OF A MATRIX

1. (a) False. Let $A = \begin{bmatrix} 0 & 1 \\ 1 & 0 \end{bmatrix}$, and suppose that A has the LU decomposition

$$\begin{bmatrix} 1 & 0 \\ a & 1 \end{bmatrix} \begin{bmatrix} b & c \\ 0 & d \end{bmatrix} = \begin{bmatrix} 0 & 1 \\ 1 & 0 \end{bmatrix}$$

for some scalars a, b, c, and d. This equation reduces to $\begin{bmatrix} b & c \\ ab & ac + d \end{bmatrix} = \begin{bmatrix} 0 & 1 \\ 1 & 0 \end{bmatrix}$, and hence $b = 0$ and $ab = 1$. But this is impossible.

(b) True

5. We apply elementary row operations to transform the given matrix into an upper triangular matrix:

$$\begin{bmatrix} -1 & 2 & 1 & -1 & 3 \\ 1 & -4 & 0 & 5 & -5 \\ -2 & 6 & -1 & -5 & 7 \\ -1 & -4 & 4 & 11 & -2 \end{bmatrix} \xrightarrow{(1)\mathbf{r}_1 + \mathbf{r}_2} \begin{bmatrix} -1 & 2 & 1 & -1 & 3 \\ 0 & -2 & 1 & 4 & -2 \\ -2 & 6 & -1 & -5 & 7 \\ -1 & -4 & 4 & 11 & -2 \end{bmatrix} \xrightarrow{(-2)\mathbf{r}_1 + \mathbf{r}_3}$$

$$\begin{bmatrix} -1 & 2 & 1 & -1 & 3 \\ 0 & -2 & 1 & 4 & -2 \\ 0 & 2 & -3 & -3 & 1 \\ -1 & -4 & 4 & 11 & -2 \end{bmatrix} \xrightarrow{(-1)\mathbf{r}_1 + \mathbf{r}_4} \begin{bmatrix} -1 & 2 & 1 & -1 & 3 \\ 0 & -2 & 1 & 4 & -2 \\ 0 & 2 & -3 & -3 & 1 \\ 0 & -6 & 3 & 12 & -5 \end{bmatrix} \xrightarrow{(1)\mathbf{r}_2 + \mathbf{r}_3}$$

$$\begin{bmatrix} -1 & 2 & 1 & -1 & 3 \\ 0 & -2 & 1 & 4 & -2 \\ 0 & 0 & -2 & 1 & -1 \\ 0 & -6 & 3 & 12 & -5 \end{bmatrix} \xrightarrow{(-3)\mathbf{r}_2 + \mathbf{r}_4} \begin{bmatrix} -1 & 2 & 1 & -1 & 3 \\ 0 & -2 & 1 & 4 & -2 \\ 0 & 0 & -2 & 1 & -1 \\ 0 & 0 & 0 & 0 & 1 \end{bmatrix} = U.$$

Since U consists of 4 rows, L is a 4×4 matrix. As in Example 3, the entries of L below the diagonal are the multipliers, and these can be obtained directly from the labels above the arrows describing the transformation of the given matrix into an upper triangular matrix. In particular, a label of the form $c\mathbf{r}_i + \mathbf{r}_j$ indicates that the (i, j)-entry of L is $-c$. Thus

$$L = \begin{bmatrix} 1 & 0 & 0 & 0 \\ -1 & 1 & 0 & 0 \\ 2 & -1 & 1 & 0 \\ 1 & 3 & 0 & 1 \end{bmatrix}.$$

9. Let

$$A = \begin{bmatrix} -1 & 2 & 1 & -1 & 3 \\ 1 & -4 & 0 & 5 & -5 \\ -2 & 6 & -1 & -5 & 7 \\ -1 & -4 & 4 & 11 & -2 \end{bmatrix} \quad \text{and} \quad \mathbf{b} = \begin{bmatrix} 7 \\ 0 \\ 8 \\ -4 \end{bmatrix}.$$

Then the system of equations can be written as $A\mathbf{x} = \mathbf{b}$. By Exercise 5, $A = LU$ is an LU decomposition of A, where

$$L = \begin{bmatrix} 1 & 0 & 0 & 0 \\ -1 & 1 & 0 & 0 \\ 2 & -1 & 1 & 0 \\ 1 & 3 & 0 & 1 \end{bmatrix} \quad \text{and} \quad U = \begin{bmatrix} -1 & 2 & 1 & -1 & 3 \\ 0 & -2 & 1 & 4 & -2 \\ 0 & 0 & -2 & 1 & -1 \\ 0 & 0 & 0 & 0 & 1 \end{bmatrix}.$$

We first solve the system $L\mathbf{y} = \mathbf{b}$, which is

$$\begin{array}{rcl} y_1 & = & 7 \\ -y_1 + y_2 & = & 7 \\ 2y_1 - y_2 + y_3 & = & 6 \\ y_1 + 3y_2 \quad + y_4 & = & 11. \end{array}$$

Clearly $y_1 = 7$. Substituting this value into the second equation, we obtain $y_2 = 0$. Continuing in this manner, we can solve for the other values to obtain

$$\mathbf{y} = \begin{bmatrix} y_1 \\ y_2 \\ y_3 \\ y_4 \end{bmatrix} = \begin{bmatrix} 7 \\ 0 \\ -8 \\ 4 \end{bmatrix}.$$

Next we solve the system $U\mathbf{x} = \mathbf{y}$, which is

$$\begin{array}{rcl} -x_1 + 2x_2 + \quad x_3 - \quad x_4 + 3x_5 & = & 7 \\ 2x_2 + \quad x_3 + 4x_4 - 2x_5 & = & 0 \\ -2x_3 + \quad x_4 - \quad x_5 & = & -8 \\ x_5 & = & 4. \end{array}$$

Now solve this system using back substitution. From the fourth equation, we have $x_5 = 4$. Substituting this value in the third equation and solving for x_3, while treating x_4 as a free variable, we obtain

$$x_3 = 2 + \frac{1}{2}x_4.$$

Similarly, we substitute the values obtained in the third and fourth equations into the second equation to solve for x_2, and we substitute the values we now have into the first equation and solve for x_1. Thus we have

$$x_2 = -3 + \frac{9}{4}x_4 \quad \text{and} \quad x_1 = 1 + 4x_4.$$

46

Therefore we obtain the general solution

$$\begin{bmatrix} x_1 \\ x_2 \\ x_3 \\ x_4 \\ x_5 \end{bmatrix} = \begin{bmatrix} 1 \\ -3 \\ 2 \\ 0 \\ 4 \end{bmatrix} + x_4 \begin{bmatrix} 4 \\ \frac{9}{4} \\ \frac{1}{2} \\ 1 \\ 0 \end{bmatrix}.$$

13. By means of elementary row operations, L can be transformed into a unit lower triangular matrix L_1 whose first column is $\mathbf{e}_1$. Additional elementary row operations can be applied to transform L_1 into a unit lower triangular matrix whose first two columns are $\mathbf{e}_1$ and $\mathbf{e}_2$. This process can be continued until L is transformed into I_n, which is in reduced row echelon form. Hence L has rank n, and thus L is invertible. So L^T is an invertible upper triangular matrix, and hence $(L^T)^{-1} = (L^{-1})^T$ is upper triangular by Exercise 11. It follows that L^{-1} is lower triangular. Let $A = L^{-1}$. For each i, the (i,i)-entry of AL is 1, and hence

$$a_{i1}l_{1i} + a_{i2}l_{2i} + \cdots + a_{ii}l_{ii} + \cdots + a_{in}l_{ni} = 1.$$

Since A and L are lower triangular, $a_{ij} = 0$ if $i < j$, and $l_{ji} = 0$ if $i > j$. Thus $a_{ij}l_{ji} = 0$ if $i \neq j$. Hence the preceding displayed equation reduces to $a_{ii}l_{ii} = 1$, and therefore $a_{ii} = 1$ because $l_{ii} = 1$.

17. We use Example 7 as a model, placing the multipliers in parentheses in appropriate matrix entries:

$$A = \begin{bmatrix} 1 & 2 & 1 & -1 \\ 2 & 4 & 1 & 1 \\ 3 & 2 & -1 & -2 \\ 2 & 5 & 3 & 0 \end{bmatrix} \xrightarrow{(-2)r_1 + r_2} \begin{bmatrix} 1 & 2 & 1 & -1 \\ (2) & 0 & -1 & 3 \\ 3 & 2 & -1 & -2 \\ 2 & 5 & 3 & 0 \end{bmatrix} \xrightarrow{(-3)r_1 + r_3} \begin{bmatrix} 1 & 2 & 1 & -1 \\ (2) & 0 & -1 & 3 \\ (3) & -4 & -4 & 1 \\ 2 & 5 & 3 & 0 \end{bmatrix}$$

$$\xrightarrow{(-2)r_1 + r_4} \begin{bmatrix} 1 & 2 & 1 & -1 \\ (2) & 0 & -1 & 3 \\ (3) & -4 & -4 & 1 \\ (2) & 1 & 1 & 2 \end{bmatrix} \xrightarrow{R(2,4)} \begin{bmatrix} 1 & 2 & 1 & -1 \\ (2) & 1 & 1 & 2 \\ (3) & -4 & -4 & 1 \\ (2) & 0 & -1 & 3 \end{bmatrix}$$

$$\xrightarrow{(4)r_2 + r_3} \begin{bmatrix} 1 & 2 & 1 & -1 \\ (2) & 1 & 1 & 2 \\ (3) & (-4) & 0 & 9 \\ (2) & 0 & -1 & 3 \end{bmatrix} \xrightarrow{R(3,4)} \begin{bmatrix} 1 & 2 & 1 & -1 \\ (2) & 1 & 1 & 2 \\ (2) & 0 & -1 & 3 \\ (3) & (-4) & 0 & 9 \end{bmatrix}.$$

The last matrix in the sequence contains the information necessary to obtain the matrices L and U of the LU decomposition of A. L is the unit lower triangular matrix whose subdiagonal entries are the same as the subdiagonal entries of the final matrix, where parentheses are removed, if necessary. U is the upper triangular matrix obtained from the final matrix in the sequence by replacing all subdiagonal entries by zeros. Thus we obtain

$$L = \begin{bmatrix} 1 & 0 & 0 & 0 \\ 2 & 1 & 0 & 0 \\ 2 & 0 & 1 & 0 \\ 3 & -4 & 0 & 1 \end{bmatrix} \quad \text{and} \quad U = \begin{bmatrix} 1 & 2 & 1 & -1 \\ 0 & 1 & 1 & 2 \\ 0 & 0 & -1 & 3 \\ 0 & 0 & 0 & 9 \end{bmatrix}.$$

Finally, we obtain P by applying the row interchanges that occur in the sequence ending in I_n. Thus

$$I_4 \xrightarrow{\ \mathbf{R}(2,4)\ } \begin{bmatrix} 1 & 0 & 0 & 0 \\ 0 & 0 & 0 & 1 \\ 0 & 0 & 1 & 0 \\ 0 & 1 & 0 & 0 \end{bmatrix} \xrightarrow{\ \mathbf{R}(3,4)\ } \begin{bmatrix} 1 & 0 & 0 & 0 \\ 0 & 0 & 0 & 1 \\ 0 & 1 & 0 & 0 \\ 0 & 0 & 1 & 0 \end{bmatrix} = P.$$

21. Let

$$A = \begin{bmatrix} 1 & 2 & 1 & -1 \\ 2 & 4 & 1 & 1 \\ 3 & 2 & -1 & -2 \\ 2 & 5 & 3 & 0 \end{bmatrix} \quad \text{and} \quad \mathbf{b} = \begin{bmatrix} 3 \\ 2 \\ -4 \\ 7 \end{bmatrix}.$$

Then the system can be written as the matrix equation $A\mathbf{x} = \mathbf{b}$. By Exercise 17, $PA = LU$, where

$$P = \begin{bmatrix} 1 & 0 & 0 & 0 \\ 0 & 0 & 0 & 1 \\ 0 & 1 & 0 & 0 \\ 0 & 0 & 1 & 0 \end{bmatrix}, \quad L = \begin{bmatrix} 1 & 0 & 0 & 0 \\ 2 & 1 & 0 & 0 \\ 2 & 0 & 1 & 0 \\ 3 & -4 & 0 & 1 \end{bmatrix}, \quad \text{and} \quad U = \begin{bmatrix} 1 & 2 & 1 & -1 \\ 0 & 1 & 1 & 2 \\ 0 & 0 & -1 & 3 \\ 0 & 0 & 0 & 9 \end{bmatrix}.$$

Since P is invertible, the system $A\mathbf{x} = \mathbf{b}$ is equivalent to

$$PA\mathbf{x} = P\mathbf{b} = \begin{bmatrix} 1 & 0 & 0 & 0 \\ 0 & 0 & 0 & 1 \\ 0 & 1 & 0 & 0 \\ 0 & 0 & 1 & 0 \end{bmatrix} \begin{bmatrix} 3 \\ 2 \\ -4 \\ 7 \end{bmatrix} = \begin{bmatrix} 3 \\ 7 \\ -2 \\ 4 \end{bmatrix} = \mathbf{b}'.$$

We can solve this system using the LU decomposition of PA given above. As in Example 4, set $\mathbf{y} = U\mathbf{x}$ and use forward substitution to solve the system $L\mathbf{y} = \mathbf{b}'$, which can be written as

$$\begin{aligned} y_1 & & & = 3 \\ 2y_1 + y_2 & & & = 7 \\ 2y_1 & + y_3 & & = 2 \\ 3y_1 - 4y_2 & & + y_4 & = -4. \end{aligned}$$

The resulting solution is

$$\mathbf{y} = \begin{bmatrix} y_1 \\ y_2 \\ y_3 \\ y_4 \end{bmatrix} = \begin{bmatrix} 3 \\ 1 \\ -4 \\ -9 \end{bmatrix}.$$

Finally, to obtain the original solution, use back substitution to solve the system $U\mathbf{x} = \mathbf{y}$, which can be written as

$$\begin{array}{rcr} x_1 + 2x_2 + x_3 - x_4 &=& 3 \\ 2x_2 + x_3 + 2x_4 &=& 1 \\ -x_3 + 3x_4 &=& -4 \\ 9x_4 &=& 9. \end{array}$$

The solution is

$$\begin{bmatrix} x_1 \\ x_2 \\ x_3 \\ x_4 \end{bmatrix} = \begin{bmatrix} -3 \\ 2 \\ 1 \\ -1 \end{bmatrix}.$$

25. Each entry of AB requires $n - 1$ additions and n multiplications for a total of $2n - 1$ flops. Since AB has mp entries, a total of $(2n - 1)mp$ flops are required to compute all the entries of AB.

29. $P = \begin{bmatrix} 0 & 1 & 0 & 0 & 0 \\ 1 & 0 & 0 & 0 & 0 \\ 0 & 0 & 1 & 0 & 0 \\ 0 & 0 & 0 & 1 & 0 \\ 0 & 0 & 0 & 0 & 1 \end{bmatrix}$, $\quad L = \begin{bmatrix} 1.0 & 0 & 0 & 0 & 0 \\ 0.0 & 1 & 0 & 0 & 0 \\ 0.5 & 2 & 1 & 0 & 0 \\ -0.5 & -1 & -3 & 1 & 0 \\ 1.5 & 7 & 9 & -9 & 1 \end{bmatrix}$, $\quad$ and

$$U = \begin{bmatrix} 2 & -2 & -1.0 & 3.0 & 4 \\ 0 & 1 & 2.0 & -1.0 & 1 \\ 0 & 0 & -1.5 & -0.5 & -2 \\ 0 & 0 & 0.0 & -1.0 & -2 \\ 0 & 0 & 0.0 & 0.0 & -9 \end{bmatrix}$$

2.6   LINEAR TRANSFORMATIONS AND MATRICES

1. (a) False, only a linear transformation has a standard matrix.
   (b) True
   (c) False, the function must also preserve vector addition.
   (d) True
   (e) False, it has size 2×3.

(f) True

(g) False, a linear transformation has this property.

(h) True

5. $T_A\left(\begin{bmatrix} 3 \\ -1 \\ 2 \end{bmatrix}\right) = A\begin{bmatrix} 3 \\ -1 \\ 2 \end{bmatrix} = \begin{bmatrix} 2 & -3 & 1 \\ 4 & 0 & -2 \end{bmatrix}\begin{bmatrix} 3 \\ -1 \\ 2 \end{bmatrix} = \begin{bmatrix} 11 \\ 8 \end{bmatrix}$

9. $T_A(\mathbf{e}_1) = \begin{bmatrix} 2 & -3 & 1 \\ 4 & 0 & -2 \end{bmatrix}\begin{bmatrix} 1 \\ 0 \\ 0 \end{bmatrix} = \begin{bmatrix} 2 \\ 4 \end{bmatrix}$. Similarly, $T_A(\mathbf{e}_2) = \begin{bmatrix} 1 \\ -2 \end{bmatrix}$.

13. $T\left(\begin{bmatrix} 16 \\ 4 \end{bmatrix}\right) = T\left(2\begin{bmatrix} 8 \\ 2 \end{bmatrix}\right) = 2T\left(\begin{bmatrix} 8 \\ 2 \end{bmatrix}\right) = 2\begin{bmatrix} 2 \\ -4 \\ 6 \end{bmatrix} = \begin{bmatrix} 4 \\ -8 \\ 12 \end{bmatrix}$

and

$$T\left(\begin{bmatrix} 4 \\ -1 \end{bmatrix}\right) = T\left(-\frac{1}{2}\begin{bmatrix} 8 \\ 2 \end{bmatrix}\right) = -\frac{1}{2}T\left(\begin{bmatrix} 8 \\ 2 \end{bmatrix}\right) = -\frac{1}{2}\begin{bmatrix} 2 \\ -4 \\ 6 \end{bmatrix} = \begin{bmatrix} -1 \\ 2 \\ -3 \end{bmatrix}$$

17. The standard matrix of T is

$$[T(\mathbf{e}_1)\ \ T(\mathbf{e}_2)] = \begin{bmatrix} 0 & 1 \\ 1 & 1 \end{bmatrix}.$$

21. The standard matrix of T is

$$[T(\mathbf{e}_1)\ \ T(\mathbf{e}_2)\ \ T(\mathbf{e}_3)] = [\mathbf{e}_1\ \ \mathbf{e}_2\ \ \mathbf{e}_3] = I_3.$$

25. $T\left(\begin{bmatrix} a \\ b \end{bmatrix}\right) = T(a\mathbf{e}_1 + b\mathbf{e}_2) = aT(\mathbf{e}_1) + bT\mathbf{e}_2 = a\begin{bmatrix} 2 \\ 3 \end{bmatrix} + b\begin{bmatrix} 4 \\ 1 \end{bmatrix} = \begin{bmatrix} 2a \\ 3a \end{bmatrix} + \begin{bmatrix} 4b \\ b \end{bmatrix} = \begin{bmatrix} 2a + 4b \\ 3a + b \end{bmatrix}$

29. T is not linear. We must show that either T does not preserve vector addition or T does not preserve scalar multiplication. For example, let $\mathbf{u} = \mathbf{e}_1$ and $\mathbf{v} = \mathbf{e}_2$. Then

$$T(\mathbf{u} + \mathbf{v}) = T(\mathbf{e}_1 + \mathbf{e}_2) = T\left(\begin{bmatrix} 1 \\ 1 \\ 0 \end{bmatrix}\right) = 1 + 1 + 0 - 1 = 1.$$

On the other hand,

$$T(\mathbf{u}) + T(\mathbf{v}) = T(\mathbf{e}_1) + T(\mathbf{e}_2) = T\left(\begin{bmatrix} 1 \\ 0 \\ 0 \end{bmatrix}\right) + T\left(\begin{bmatrix} 0 \\ 1 \\ 0 \end{bmatrix}\right)$$

$$= (1 + 0 + 0 - 1) + (0 + 1 + 0 - 1) = 0.$$

So $T(\mathbf{u}+\mathbf{v}) \neq T(\mathbf{u})+T(\mathbf{v})$ for the given vectors. Therefore T does not preserve vector addition and hence is not linear.

ALTERNATE PROOF. Let $c = 4$ and $\mathbf{u} = \mathbf{e}_1$. Then

$$T(4\mathbf{u}) = T(4\mathbf{e}_1) = T\left(\begin{bmatrix} 4 \\ 0 \\ 0 \end{bmatrix}\right) = 4 + 0 + 0 - 1 = 3.$$

On the other hand,

$$4T(\mathbf{u}) = 4T(\mathbf{e}_1) = 4T\left(\begin{bmatrix} 1 \\ 0 \\ 0 \end{bmatrix}\right) = 4(1 + 0 + 0 - 1) = 0.$$

So $T(4\mathbf{u}) \neq 4T(\mathbf{u})$ and hence T does not preserve scalar multiplication. Therefore T is not linear.

COMMENT. For this example, we can also show that T is not linear by noting that $T(\mathbf{0}) = 0 + 0 + 0 - 1 = -1 \neq 0$. By Theorem 2.9(a), we conclude that T is not linear.

33. T is not linear. Let $\mathbf{u} = \begin{bmatrix} \frac{\pi}{2} \\ 0 \end{bmatrix}$. Then

$$T(2\mathbf{u}) = T\left(\begin{bmatrix} \pi \\ 0 \end{bmatrix}\right) = \begin{bmatrix} \sin \pi \\ 0 \end{bmatrix} = \begin{bmatrix} 0 \\ 0 \end{bmatrix} = \mathbf{0},$$

but

$$2T(\mathbf{u}) = 2T\left(\begin{bmatrix} \frac{\pi}{2} \\ 0 \end{bmatrix}\right) = 2\begin{bmatrix} \sin \frac{\pi}{2} \\ 0 \end{bmatrix} = 2\begin{bmatrix} 1 \\ 0 \end{bmatrix} = \begin{bmatrix} 2 \\ 0 \end{bmatrix}.$$

So $T(2\mathbf{u}) \neq 2T(\mathbf{u})$, and hence T does not preserve scalar multiplication.

37. We must show that the transformation cT preserves vector addition and scalar multiplication. Let $\mathbf{u}$ and $\mathbf{v}$ be in $\mathcal{R}^n$. Then because T is linear, we have

$$(cT)(\mathbf{u} + \mathbf{v}) = cT(\mathbf{u} + \mathbf{v}) = c(T(\mathbf{u}) + T(\mathbf{v})) = cT(\mathbf{u}) + cT(\mathbf{v}).$$

Also

$$(cT)(\mathbf{u}) + (cT)(\mathbf{v}) = cT(\mathbf{u}) + cT(\mathbf{v}).$$

So cT preserves vector addition. Now suppose k is a scalar. Then because T is linear, we have

$$(cT)(k\mathbf{u}) = c(T(k\mathbf{u})) = c(kT(\mathbf{u})) = ckT(\mathbf{u}).$$

Also

$$k((cT)(\mathbf{u})) = k(cT(\mathbf{u})) = kcT(\mathbf{u}) = ckT(\mathbf{u}).$$

So cT preserves scalar multiplication. Hence cT is linear.

41. By Theorem 2.11, there exists a matrix A such that $T(\mathbf{v}) = A\mathbf{v}$ for all $\mathbf{v}$ in $\mathcal{R}^2$. Let $A = \begin{bmatrix} a & b \\ c & d \end{bmatrix}$. Then

$$T\left(\begin{bmatrix} x_1 \\ x_2 \end{bmatrix}\right) = \begin{bmatrix} a & b \\ c & d \end{bmatrix}\begin{bmatrix} x_1 \\ x_2 \end{bmatrix} = \begin{bmatrix} ax_1 + bx_2 \\ cx_1 + dx_2 \end{bmatrix}.$$

45. (a) Because it is given that T is linear, it follows from Theorem 2.11 that T is a matrix transformation.

ALTERNATE PROOF. Let $A = \begin{bmatrix} -1 & 0 \\ 0 & 1 \end{bmatrix}$. Then

$$T_A\left(\begin{bmatrix} x_1 \\ x_2 \end{bmatrix}\right) = \begin{bmatrix} -1 & 0 \\ 0 & 1 \end{bmatrix}\begin{bmatrix} x_1 \\ x_2 \end{bmatrix} = \begin{bmatrix} -x_1 \\ x_2 \end{bmatrix} = T\left(\begin{bmatrix} x_1 \\ x_2 \end{bmatrix}\right),$$

and hence $T = T_A$.

(b) For any vector $\mathbf{v} = \begin{bmatrix} v_1 \\ v_2 \end{bmatrix}$ in $\mathcal{R}^2$,

$$T\left(\begin{bmatrix} -v_1 \\ v_2 \end{bmatrix}\right) = \begin{bmatrix} -(-v_1) \\ v_2 \end{bmatrix} = \begin{bmatrix} v_1 \\ v_2 \end{bmatrix} = \mathbf{v},$$

and hence the range of T is $\mathcal{R}^2$.

49. Because T is linear, we have $T(\mathbf{u}) = T(\mathbf{v})$ if and only if $T(\mathbf{u}) - T(\mathbf{v}) = \mathbf{0}$. This equation holds if and only if $T(\mathbf{u} - \mathbf{v}) = \mathbf{0}$.

53. A vector $\mathbf{v}$ is in the range of T if and only if $\mathbf{v} = T(\mathbf{u}) = A\mathbf{u}$ for some $\mathbf{u}$ in $\mathcal{R}^n$, which is true if and only if $\mathbf{v}$ is in the span of the columns of A.

57. The given vector $\mathbf{v} = \begin{bmatrix} 2 \\ -1 \\ 0 \\ 3 \end{bmatrix}$ is in the range of T if and only there is a vector $\mathbf{u}$ such that $T(\mathbf{u}) = \mathbf{v}$. If A is the standard matrix of T, then this condition is equivalent to the system $A\mathbf{x} = \mathbf{v}$ being consistent, where

$$A = \begin{bmatrix} 1 & 1 & 1 & 2 \\ 1 & 2 & -3 & 4 \\ 0 & 1 & 0 & 2 \\ 1 & 5 & -1 & 0 \end{bmatrix}.$$

If we solve this system, we obtain

$$\mathbf{u} = \frac{1}{4}\begin{bmatrix} 5 \\ 2 \\ 3 \\ -1 \end{bmatrix}.$$

So $T(\mathbf{u}) = \mathbf{v}$, and therefore $\mathbf{v}$ is in the range of T.

Alternatively, we can show that the reduced row echelon form of A is I_4 and conclude from Theorem 2.6 that the system $A\mathbf{x} = \mathbf{b}$ is consistent for every $\mathbf{b}$ in $\mathcal{R}^4$.

2.7 COMPOSITION AND INVERTIBILITY OF LINEAR TRANSFORMATIONS

1. (a) True
   (b) False, the columns *always* span the range.
   (c) False, we need that the columns span the codomain. The matrix
   $$A = \begin{bmatrix} 1 & 0 \\ 0 & 1 \\ 0 & 0 \end{bmatrix}$$
   has linearly independent columns, but the vector $\begin{bmatrix} 0 \\ 0 \\ 1 \end{bmatrix}$ is not in the range of T_A.
   (d) True
   (e) True
   (f) True
   (g) False, consider the 3×2 matrix A in (c). The rank of A is 2, and so T_A is one-to-one, but T_A is not onto because we need the rank to be 3.
   (h) True

5. By the boxed result on page 161, the columns of the standard matrix of T span the range of T. So the desired set is
   $$\{T(\mathbf{e}_1), T(\mathbf{e}_2)\} = \left\{ \begin{bmatrix} 0 \\ 2 \\ 1 \end{bmatrix}, \begin{bmatrix} 3 \\ -1 \\ 1 \end{bmatrix} \right\}.$$

9. By the boxed result on page 161, the columns of the standard matrix of T span the range of T. So the described set is
   $$\{T(\mathbf{e}_1), T(\mathbf{e}_2), T(\mathbf{e}_3)\} = \{\mathbf{0}, \mathbf{0}, \mathbf{0}\} = \{\mathbf{0}\} = \left\{ \begin{bmatrix} 0 \\ 0 \end{bmatrix} \right\}.$$

13. The null space of T is the solution set of $A\mathbf{x} = \mathbf{0}$, where
    $$A = \begin{bmatrix} 0 & 1 \\ 1 & 1 \end{bmatrix}$$
    is the standard matrix of T. Thus the general solution of $A\mathbf{x} = \mathbf{0}$ is
    $$x_1 = 0$$

$$x_2 = 0.$$

So a spanning set is $\{\mathbf{0}\}$. By Theorem 2.13, we have that T is one-to-one.

17. The null space of T is the solution set of $A\mathbf{x} = \mathbf{0}$, where

$$A = \begin{bmatrix} 1 & 2 & 1 \\ 1 & 3 & 2 \\ 2 & 5 & 3 \end{bmatrix}$$

is the standard matrix of T. The general solution of $A\mathbf{x} = \mathbf{0}$ is

$$\begin{aligned} x_1 &= x_3 \\ x_2 &= -x_3 \\ x_3 & \quad \text{free,} \end{aligned}$$

or

$$\begin{bmatrix} x_1 \\ x_2 \\ x_3 \end{bmatrix} = \begin{bmatrix} x_3 \\ -x_3 \\ x_3 \end{bmatrix} = x_3 \begin{bmatrix} 1 \\ -1 \\ 1 \end{bmatrix}.$$

So a spanning set for the null space of T is

$$\left\{ \begin{bmatrix} 1 \\ -1 \\ 1 \end{bmatrix} \right\}.$$

By Theorem 2.13, we have that T is not one-to-one.

21. The null space of T is the solution set of $A\mathbf{x} = \mathbf{0}$, where

$$A = \begin{bmatrix} 1 & 0 \\ 0 & 0 \end{bmatrix}$$

is the standard matrix of T. The general solution of $A\mathbf{x} = \mathbf{0}$ is

$$\begin{aligned} x_1 &= 0 \\ x_2 & \quad \text{free,} \end{aligned}$$

or

$$\begin{bmatrix} x_1 \\ x_2 \end{bmatrix} = \begin{bmatrix} 0 \\ x_2 \end{bmatrix} = x_2 \begin{bmatrix} 0 \\ 1 \end{bmatrix} = x_2 \mathbf{e}_2.$$

So a spanning set for the null space of T is $\{\mathbf{e}_2\}$. By Theorem 2.13, we have that T is not one-to-one.

25. The standard matrix of T is

$$[T(\mathbf{e}_1) \ T(\mathbf{e}_2)] = \begin{bmatrix} 2 & 3 \\ 4 & 5 \end{bmatrix}.$$

The reduced row echelon form of this matrix is

$$\begin{bmatrix} 1 & 0 \\ 0 & 1 \end{bmatrix},$$

which has rank 2. So by Theorem 2.13, T is one-to-one.

29. The standard matrix of T is $A = \begin{bmatrix} 2 & 3 \\ 4 & 5 \end{bmatrix}$. Clearly rank $A = 2$; so by Theorem 2.12 we have that T is onto.

33. (a) The null space of T is $\{\mathbf{0}\}$ because the only vector that is rotated into $\mathbf{0}$ is the zero vector.
   (b) Yes, by Theorem 2.3.
   (c) $\mathcal{R}^2$. For any vector $\mathbf{v}$ in $\mathcal{R}^2$, let $\mathbf{u}$ be the vector formed by rotating $\mathbf{v}$ clockwise by $90°$. Clearly $T(\mathbf{u}) = \mathbf{v}$. So every vector in $\mathcal{R}^2$ is in the range of T.
   (d) Yes, by (c).

37. (a) The null space of T is Span $\{\mathbf{e}_3\}$ because the only vectors that are projected onto $\{\mathbf{0}\}$ are the vectors along the z-axis.
   (b) No, because of (a) and Theorem 2.13.
   (c) Because T is the projection onto the xy-plane, every vector has an image that is in the xy-plane, that is, the range of T is Span $\{\mathbf{e}_1, \mathbf{e}_2\}$.
   (d) No, by (c), the range of T is not $\mathcal{R}^3$.

41. The domain and codomain are both $\mathcal{R}^2$. Also

$$(UT)\left(\begin{bmatrix} x_1 \\ x_2 \end{bmatrix}\right) = U\left(T\left(\begin{bmatrix} x_1 \\ x_2 \end{bmatrix}\right)\right) = U\left(\begin{bmatrix} x_1 + x_2 \\ x_1 - 3x_2 \\ 4x_1 \end{bmatrix}\right)$$

$$= \begin{bmatrix} (x_1 + x_2) - (x_1 - 3x_2) + 4(4x_1) \\ (x_1 + x_2) + 3(x_1 - 3x_2) \end{bmatrix} = \begin{bmatrix} 16x_1 + 4x_2 \\ 4x_1 - 8x_2 \end{bmatrix}.$$

45. The domain and codomain are $\mathcal{R}^3$. The rule is

$$(TU)\left(\begin{bmatrix} x_1 \\ x_2 \\ x_3 \end{bmatrix}\right) = \begin{bmatrix} 2x_1 + 2x_2 + 4x_3 \\ -2x_1 - 10x_2 + 4x_3 \\ 4x_1 - 4x_2 + 16x_3 \end{bmatrix}.$$

49. We first need to find the rule for UT.

$$(UT)\left(\begin{bmatrix} x_1 \\ x_2 \end{bmatrix}\right) = U\left(T\left(\begin{bmatrix} x_1 \\ x_2 \end{bmatrix}\right)\right) = U\left(\begin{bmatrix} x_1 + 2x_2 \\ 3x_1 - x_2 \end{bmatrix}\right)$$

$$= \begin{bmatrix} 2(x_1 + 2x_2) - (3x_1 - x_2) \\ 5(3x_1 - x_2) \end{bmatrix} = \begin{bmatrix} -x_1 + 5x_2 \\ 15x_1 - 5x_2 \end{bmatrix}.$$

So the standard matrix of UT is

$$[(UT)(e_1) \ (UT)(e_2)] = \begin{bmatrix} -1 & 5 \\ 15 & -5 \end{bmatrix}.$$

53. We first need to find the rule for TU.

$$(TU)\left(\begin{bmatrix} x_1 \\ x_2 \end{bmatrix}\right) = T\left(U\left(\begin{bmatrix} x_1 \\ x_2 \end{bmatrix}\right)\right) = T\left(\begin{bmatrix} 2x_1 - x_2 \\ 5x_2 \end{bmatrix}\right)$$

$$= \begin{bmatrix} (2x_1 - x_2) + 2(5x_2) \\ 3(2x_1 - x_2) - (5x_2) \end{bmatrix} = \begin{bmatrix} 2x_1 + 9x_2 \\ 6x_1 - 8x_2 \end{bmatrix}.$$

So the standard matrix of TU is

$$[(TU)(e_1) \ (TU)(e_2)] = \begin{bmatrix} 2 & 9 \\ 6 & -8 \end{bmatrix}.$$

57. Let A be the standard matrix of T. Then by Theorem 2.14, $T^{-1} = T_{A^{-1}}$. We have that

$$A = \begin{bmatrix} -1 & 1 & 3 \\ 2 & 0 & -1 \\ -1 & 2 & 5 \end{bmatrix} \quad \text{and} \quad A^{-1} = \begin{bmatrix} 2 & 1 & -1 \\ -9 & -2 & 5 \\ 4 & 1 & -2 \end{bmatrix}.$$

So

$$T^{-1}\left(\begin{bmatrix} x_1 \\ x_2 \\ x_3 \end{bmatrix}\right) = \begin{bmatrix} 2x_1 + x_2 - x_3 \\ -9x_1 - 2x_2 + 5x_3 \\ 4x_1 + x_2 - 2x_3 \end{bmatrix}.$$

61. If T is the reflection about the x-axis, then

$$T\left(\begin{bmatrix} x_1 \\ x_2 \end{bmatrix}\right) = \begin{bmatrix} x_1 \\ -x_2 \end{bmatrix}.$$

So

$$(TT)\left(\begin{bmatrix} x_1 \\ x_2 \end{bmatrix}\right) = T\left(T\left(\begin{bmatrix} x_1 \\ x_2 \end{bmatrix}\right)\right) = T\left(\begin{bmatrix} x_1 \\ -x_2 \end{bmatrix}\right)$$

$$= T\left(\begin{bmatrix} x_1 \\ -(-x_2) \end{bmatrix}\right) = \begin{bmatrix} x_1 \\ x_2 \end{bmatrix} = I\left(\begin{bmatrix} x_1 \\ x_2 \end{bmatrix}\right).$$

So $TT = I$.

65. (a) Suppose

$$a_1 T(\mathbf{v}_1) + a_2 T(\mathbf{v}_2) + \cdots + a_k T(\mathbf{v}_k) = \mathbf{0}$$

for some scalars $a_1, a_2, \ldots, a_k$. Because T is linear, we have

$$T(a_1 \mathbf{v}_1 + a_2 \mathbf{v}_2 + \cdots + a_k \mathbf{v}_k) = \mathbf{0}.$$

So $a_1 \mathbf{v}_1 + a_2 \mathbf{v}_2 + \cdots + a_k \mathbf{v}_k$ is in the null space of T. Therefore, by Theorem 2.13, $a_1 \mathbf{v}_1 + a_2 \mathbf{v}_2 + \cdots + a_k \mathbf{v}_k = \mathbf{0}$. Because $\{\mathbf{v}_1, \mathbf{v}_2, \ldots, \mathbf{v}_k\}$ is linearly independent, we have $a_1 = a_2 = \cdots = a_k = 0$. Hence $\{T(\mathbf{v}_1), T(\mathbf{v}_2), \ldots, T(\mathbf{v}_k)\}$ is linearly independent.

(b) Let T be the projection on the x-axis. Then $\{\mathbf{e}_2\}$ is linearly independent, but $\{T(\mathbf{e}_1)\} = \{\mathbf{0}\}$ is not.

CHAPTER 2 REVIEW

1. (a) True

(b) False, consider $\begin{bmatrix} 1 & 2 \\ 2 & 4 \end{bmatrix}$.

(c) False, the product of a 2×2 and 3×3 matrix is not defined.

(d) True

(e) False, see the boxed result on page 114.

(f) False, consider $I_2 + I_2 = 2I_2 \neq O$.

(g) True

(h) True

(i) False, $O\mathbf{x} = \mathbf{0}$ is consistent for $\mathbf{b} = \mathbf{0}$, but O is not invertible.

(j) True

(k) False, the null space is contained in the domain.

(l) True

(m) False, let T be the projection on the x-axis. Then $\{\mathbf{e}_2\}$ is linearly independent but $\{T(\mathbf{e}_2)\} = \{\mathbf{0}\}$ is not linearly independent.

(n) True

(o) True

(p) False, the transformation $T_A\left(\begin{bmatrix} x_1 \\ x_2 \end{bmatrix}\right) = \begin{bmatrix} x_1 \\ x_2 \\ 0 \end{bmatrix}$ is one-to-one, but **not onto**.

(q) True

(r) False, the null space consists exactly of the zero vector.

(s) False, the columns of its standard matrix span its codomain.

(t) True

5. We have

$$ABA = A(BA) = \begin{bmatrix} 2 & 1 \\ 4 & -1 \end{bmatrix} \left(\begin{bmatrix} 2 & 3 \\ 4 & 6 \end{bmatrix} \begin{bmatrix} 2 & 1 \\ 4 & -1 \end{bmatrix} \right)$$

$$= \begin{bmatrix} 2 & 1 \\ 4 & -1 \end{bmatrix} \begin{bmatrix} 16 & -1 \\ 32 & -2 \end{bmatrix} = \begin{bmatrix} 64 & -4 \\ 32 & -2 \end{bmatrix}.$$

9. The product $\mathbf{v}A$ is undefined because in $(1 \times 3)(2 \times 2)$ the inner dimensions are unequal.

13. We have

$$AC^T\mathbf{u} = \begin{bmatrix} 2 & 1 \\ 4 & -1 \end{bmatrix} \left(\begin{bmatrix} 2 & 3 & 0 \\ -1 & 5 & 1 \end{bmatrix} \begin{bmatrix} 3 \\ 2 \\ -1 \end{bmatrix} \right)$$

$$= \begin{bmatrix} 2 & 1 \\ 4 & -1 \end{bmatrix} \begin{bmatrix} 12 \\ 6 \end{bmatrix} = \begin{bmatrix} 30 \\ 42 \end{bmatrix}.$$

17. We have

$$[I_2 \mid -I_2] \begin{bmatrix} 1 \\ 3 \\ \hline -7 \\ -4 \end{bmatrix} = I_2 \begin{bmatrix} 1 \\ 3 \end{bmatrix} + (-I_2) \begin{bmatrix} -7 \\ -4 \end{bmatrix}$$

$$= \begin{bmatrix} 1 \\ 3 \end{bmatrix} + \begin{bmatrix} 7 \\ 4 \end{bmatrix} = \begin{bmatrix} 8 \\ 7 \end{bmatrix}.$$

21. By item 1 on page 95, the first column of AB is $A\mathbf{b}_1 = A\mathbf{0} = \mathbf{0}$.

25. We use the linear correspondence property. Because

$$\mathbf{r}_2 = 2\mathbf{r}_1 \qquad \text{and} \qquad \mathbf{r}_5 = -2\mathbf{r}_1 + 3\mathbf{r}_3 + \mathbf{r}_4,$$

we have

$$\mathbf{a}_2 = 2\mathbf{a}_1 = 2 \begin{bmatrix} 3 \\ 5 \\ 2 \end{bmatrix} = \begin{bmatrix} 6 \\ 10 \\ 4 \end{bmatrix}$$

and

$$\mathbf{a}_5 = -2\mathbf{a}_1 + 3\mathbf{a}_3 + \mathbf{a}_4$$

$$= -2 \begin{bmatrix} 3 \\ 5 \\ 2 \end{bmatrix} + 3 \begin{bmatrix} 2 \\ 0 \\ -1 \end{bmatrix} + \begin{bmatrix} 2 \\ -1 \\ 3 \end{bmatrix} = \begin{bmatrix} 2 \\ -11 \\ -4 \end{bmatrix}.$$

So

$$A = \begin{bmatrix} 3 & 6 & 2 & 2 & 2 \\ 5 & 10 & 0 & -1 & -11 \\ 2 & 4 & -1 & 3 & -4 \end{bmatrix}.$$

29. $T_B\left(\begin{bmatrix} 4 \\ 2 \end{bmatrix}\right) = \begin{bmatrix} 4 & 2 \\ 1 & -3 \\ 0 & 1 \end{bmatrix} \begin{bmatrix} 4 \\ 2 \end{bmatrix} = \begin{bmatrix} 20 \\ -2 \\ 2 \end{bmatrix}$

33. The standard matrix of T is $A = [T(\mathbf{e}_1)\ T(\mathbf{e}_2)]$. Now

$$T(\mathbf{e}_1) = 2\mathbf{e}_1 + U(\mathbf{e}_1) = 2\begin{bmatrix} 1 \\ 0 \end{bmatrix} + U\left(\begin{bmatrix} 1 \\ 0 \end{bmatrix}\right) = \begin{bmatrix} 2 \\ 0 \end{bmatrix} + \begin{bmatrix} 2 \\ 3 \end{bmatrix} = \begin{bmatrix} 4 \\ 3 \end{bmatrix}.$$

Also

$$T(\mathbf{e}_2) = 2\mathbf{e}_2 + U(\mathbf{e}_2) = 2\begin{bmatrix} 0 \\ 1 \end{bmatrix} + U\left(\begin{bmatrix} 0 \\ 1 \end{bmatrix}\right) = \begin{bmatrix} 0 \\ 2 \end{bmatrix} + \begin{bmatrix} 1 \\ 0 \end{bmatrix} = \begin{bmatrix} 1 \\ 2 \end{bmatrix}.$$

So

$$A = \begin{bmatrix} 4 & 1 \\ 3 & 2 \end{bmatrix}.$$

37. Let $\mathbf{u}$ and $\mathbf{v}$ be vectors in $\mathcal{R}^3$. Then

$$T(\mathbf{u} + \mathbf{v}) = T\left(\begin{bmatrix} u_1 + v_1 \\ u_2 + v_2 \\ u_3 + v_3 \end{bmatrix}\right) = \begin{bmatrix} (u_1 + v_1) + (u_2 + v_2) \\ u_3 + v_3 \end{bmatrix}$$

$$= \begin{bmatrix} u_1 + u_2 \\ u_3 \end{bmatrix} + \begin{bmatrix} v_1 + v_2 \\ v_3 \end{bmatrix} = T(\mathbf{u}) + T(\mathbf{v}).$$

So T preserves vector addition.

Let c be any scalar. Then

$$T(c\mathbf{u}) = T\left(\begin{bmatrix} cu_1 \\ cu_2 \\ cu_3 \end{bmatrix}\right) = \begin{bmatrix} cu_1 + cu_2 \\ cu_3 \end{bmatrix}$$

$$= c\begin{bmatrix} u_1 + u_2 \\ u_3 \end{bmatrix} = cT(\mathbf{u}).$$

Thus T preserves scalar multiplication. So T is linear.
ALTERNATE PROOF. Let

$$A = \begin{bmatrix} 1 & 1 & 0 \\ 0 & 0 & 1 \end{bmatrix}.$$

Then

$$T_A\left(\begin{bmatrix} x_1 \\ x_2 \\ x_3 \end{bmatrix}\right) = \begin{bmatrix} 1 & 1 & 0 \\ 0 & 0 & 1 \end{bmatrix} \begin{bmatrix} x_1 \\ x_2 \\ x_3 \end{bmatrix} = \begin{bmatrix} x_1 + x_2 \\ x_3 \end{bmatrix}.$$

So $T = T_A$, and hence T is linear because T_A is linear.

41. The null space of T is the solution set to $A\mathbf{x} = \mathbf{0}$, where A is the standard matrix of T. The general solution of $A\mathbf{x} = \mathbf{0}$ is

$$\begin{aligned} x_1 &= -2x_3 \\ x_2 &= x_3 \\ x_3 &\quad \text{free.} \end{aligned}$$

Thus

$$\begin{bmatrix} x_1 \\ x_2 \\ x_3 \end{bmatrix} = \begin{bmatrix} -2x_3 \\ x_3 \\ x_3 \end{bmatrix} = x_3 \begin{bmatrix} -2 \\ 1 \\ 1 \end{bmatrix}.$$

So the spanning set is $\left\{ \begin{bmatrix} -2 \\ 1 \\ 1 \end{bmatrix} \right\}$. Thus T is not one-to-one by Theorem 2.13.

45. The standard matrix of T is

$$[T(\mathbf{e}_1) \ \ T(\mathbf{e}_2)] = \begin{bmatrix} 3 & -1 \\ 0 & 1 \\ 1 & 1 \end{bmatrix}.$$

The rank of this matrix is 2, so T is not onto by Theorem 2.12.

49. We have

$$BA = \begin{bmatrix} 3 & -1 \\ 0 & 1 \\ 1 & 1 \end{bmatrix} \begin{bmatrix} 2 & 0 & 1 \\ 1 & 1 & -1 \end{bmatrix} = \begin{bmatrix} 5 & -1 & 4 \\ 1 & 1 & -1 \\ 3 & 1 & 0 \end{bmatrix}.$$

We need to find the rule for UT.

$$(UT)\left(\begin{bmatrix} x_1 \\ x_2 \\ x_3 \end{bmatrix}\right) = U\left(T\left(\begin{bmatrix} x_1 \\ x_2 \\ x_3 \end{bmatrix}\right)\right) = U\left(\begin{bmatrix} 2x_1 + x_3 \\ x_1 + x_2 - x_3 \end{bmatrix}\right)$$

$$= \begin{bmatrix} 3(2x_1 + x_3) - (x_1 + x_2 - x_3) \\ x_1 + x_2 - x_3 \\ (2x_1 + x_3) + (x_1 + x_2 - x_3) \end{bmatrix} = \begin{bmatrix} 5x_1 - x_2 + 4x_3 \\ x_1 + x_2 - x_3 \\ 3x_1 + x_2 \end{bmatrix}.$$

The standard matrix of UT is

$$[(UT)(\mathbf{e}_1) \ \ (UT)(\mathbf{e}_2) \ \ (UT)(\mathbf{e}_3)] = \begin{bmatrix} 5 & -1 & 4 \\ 1 & 1 & -1 \\ 3 & 1 & 0 \end{bmatrix} = BA.$$

53. The standard matrix A of T is

$$A = [T(\mathbf{e}_1) \ T(\mathbf{e}_2)] = \begin{bmatrix} 1 & 2 \\ -1 & 3 \end{bmatrix}.$$

Note that

$$A^{-1} = \frac{1}{5} \begin{bmatrix} 3 & -2 \\ 1 & 1 \end{bmatrix}.$$

Thus, by Theorem 2.14(b), we have

$$T^{-1}\left(\begin{bmatrix} x_1 \\ x_2 \end{bmatrix}\right) = T_{A^{-1}}\left(\begin{bmatrix} x_1 \\ x_2 \end{bmatrix}\right) = A^{-1}\begin{bmatrix} x_1 \\ x_2 \end{bmatrix}$$

$$= \frac{1}{5} \begin{bmatrix} 3 & -2 \\ 1 & 1 \end{bmatrix} \begin{bmatrix} x_1 \\ x_2 \end{bmatrix} = \frac{1}{5} \begin{bmatrix} 3x_1 - 2x_2 \\ x_1 + x_2 \end{bmatrix}.$$

Chapter 3

Determinants

3.1 COFACTOR EXPANSION

1. **(a)** False, det $\begin{bmatrix} a & b \\ c & d \end{bmatrix} = ad - bc$.

   (b) False, the (i, j)-cofactor of A equals $(-1)^{i+j}$ times the determinant of the $(n-1) \times (n-1)$ matrix obtained by deleting row i and column j from A.

   (c) True

   (d) False, the determinant of an upper triangular or a lower triangular square matrix equals the *product* of its diagonal entries.

   (e) False, the area of the parallelogram determined by $\mathbf{u}$ and $\mathbf{v}$ is $|\det [\mathbf{u} \ \mathbf{v}]|$.

5. $\det \begin{bmatrix} -5 & -6 \\ 10 & 12 \end{bmatrix} = (-5)(12) - (-6)(10) = -60 + 60 = 0$

9. The $(3, 1)$-cofactor of A is

$$(-1)^{3+1} \det \begin{bmatrix} -2 & 4 \\ 6 & 3 \end{bmatrix} = 1[(-2)(3) - 4(6)] = 1(-30) = -30.$$

13. The cofactor expansion along the third row is

$$0(-1)^{3+1} \det \begin{bmatrix} -2 & 2 \\ -1 & 3 \end{bmatrix} + 1(-1)^{3+2} \det \begin{bmatrix} 1 & 2 \\ 2 & 3 \end{bmatrix} + (-1)(-1)^{3+3} \det \begin{bmatrix} 1 & -2 \\ 2 & -1 \end{bmatrix}$$

$$= 0 + (-1)[1(3) - 2(2)] + (-1)[1(-1) - (-2)(2)]$$

$$= -2.$$

17. The cofactor expansion along the second row is

$$0 + (-1)(-1)^{2+2} \det \begin{bmatrix} 1 & 1 & -1 \\ 4 & 2 & -1 \\ 0 & 0 & -2 \end{bmatrix} + 0 + 1(-1)^{2+4} \det \begin{bmatrix} 1 & 2 & 1 \\ 4 & -3 & 2 \\ 0 & 3 & 0 \end{bmatrix}$$

$$= (-1)(-2)(-1)^{3+3} \det \begin{bmatrix} 1 & 1 \\ 4 & 2 \end{bmatrix} + 1(3)(-1)^{3+2} \det \begin{bmatrix} 1 & 1 \\ 4 & 2 \end{bmatrix}$$

$$= 2[1(2) - 1(4)] - 3[1(2) - 1(4)]$$

$$= 2(-2) - 3(-2)$$

$$= 2.$$

21. We have

$$\det \begin{bmatrix} -6 & 0 & 0 \\ 7 & -3 & 2 \\ 2 & 9 & 4 \end{bmatrix} = -6(-1)^{1+1} \det \begin{bmatrix} -3 & 2 \\ 9 & 4 \end{bmatrix}$$

$$= -6[(-3)4 - 2(9)]$$

$$= 180.$$

25. We have

$$\det \begin{bmatrix} -2 & -1 & -5 & 1 \\ 0 & 0 & 0 & 4 \\ 0 & -2 & 0 & 5 \\ 3 & 1 & 6 & 2 \end{bmatrix} = 4(-1)^{2+4} \det \begin{bmatrix} -2 & -1 & -5 \\ 0 & -2 & 0 \\ 3 & 1 & 6 \end{bmatrix}$$

$$= 4(-2)(-1)^{2+2} \det \begin{bmatrix} -2 & -5 \\ 3 & 6 \end{bmatrix}$$

$$= -8[-2(6) - (-5)(3)]$$

$$= -24.$$

29. The area of the parallelogram determined by **u** and **v** is

$$|\det [\mathbf{u}\ \mathbf{v}]| = \left| \det \begin{bmatrix} 6 & 3 \\ 4 & 2 \end{bmatrix} \right| = |6(2) - 3(4)| = |0| = 0.$$

33. We have

$$\det A_\theta = \det \begin{bmatrix} \cos\theta & -\sin\theta \\ \sin\theta & \cos\theta \end{bmatrix} = \cos^2\theta - (-\sin^2\theta)$$

$$= \cos^2\theta + \sin^2\theta = 1.$$

37. Let $A = \begin{bmatrix} a & b \\ c & d \end{bmatrix}$. Then

$$\det A^T = \det \begin{bmatrix} a & c \\ b & d \end{bmatrix} = ad - cb = \det \begin{bmatrix} a & b \\ c & d \end{bmatrix} = \det A.$$

63

41. We have $\det E = k$ and $\det A = ad - bc$; so $(\det E)(\det A) = k(ad - bc)$. Also

$$EA = \begin{bmatrix} a & b \\ kc & kd \end{bmatrix}.$$

Thus

$$\det EA = a(kd) - b(kc) = k(ad - bc) = (\det E)(\det A).$$

45. We have

$$\det \begin{bmatrix} a & b \\ c + kp & d + kq \end{bmatrix} = a(d + kq) - b(c + kp)$$

$$= ad + akq - bc - bkp$$

and

$$\det \begin{bmatrix} a & b \\ c & d \end{bmatrix} + k \det \begin{bmatrix} a & b \\ p & q \end{bmatrix} = (ad - bc) + k(aq - bp)$$

$$= ad - bc + akq - bkq$$

$$= ad + akq - bc - bkp$$

$$= a(d + kq) - b(c + kp)$$

$$= \det \begin{bmatrix} a & b \\ c + kp & d + kq \end{bmatrix}$$

49. (c) no

3.2   PROPERTIES OF DETERMINANTS

1. (a) False, $\det \begin{bmatrix} 1 & 2 \\ 3 & 4 \end{bmatrix} \neq 1 \cdot 4$.

   (b) True
   (c) False, multiplying a row of a square matrix by a scalar c changes the determinant by a factor of c.
   (d) True
   (e) False, consider $A = [\mathbf{e}_1 \ \mathbf{0}]$ and $B = [\mathbf{0} \ \mathbf{e}_2]$.
   (f) True
   (g) False, if A is an invertible matrix, then $\det A \neq 0$.
   (h) False, for any square matrix A, $\det A^T = \det A$.
   (i) True
   (j) False, the determinant of $2I_2$ is 4, but its reduced row echelon form is I_2.

3. The cofactor expansion along the second column yields

$$- 1(-1)^{1+2} \det \begin{bmatrix} 1 & -2 \\ -1 & 1 \end{bmatrix} + 4(-1)^{2+2} \det \begin{bmatrix} 2 & 3 \\ -1 & 1 \end{bmatrix} + 0(-1)^{3+2} \det \begin{bmatrix} 2 & 3 \\ -1 & 2 \end{bmatrix}$$

$$= 1[1(1) - (-2)(-1)] + 4[2(1) - 3(-1)] + 0[2(2) - 3(-1)]$$
$$= 1(-1) + 4(5) + 0(7)$$
$$= 19.$$

5. $\det \begin{bmatrix} 0 & 0 & 5 \\ 0 & 3 & 7 \\ 4 & -1 & -2 \end{bmatrix} = - \det \begin{bmatrix} 4 & -1 & 2 \\ 0 & 3 & 7 \\ 0 & 0 & 5 \end{bmatrix} = -(4)(3)(5) = -60$

9. We have

$$\det \begin{bmatrix} 3 & -2 & 1 \\ 0 & 0 & 5 \\ -9 & 4 & 2 \end{bmatrix} = - \det \begin{bmatrix} 3 & -2 & 1 \\ -9 & 4 & 2 \\ 0 & 0 & 5 \end{bmatrix} = - \det \begin{bmatrix} 3 & -2 & 1 \\ 0 & -2 & 5 \\ 0 & 0 & 5 \end{bmatrix}$$

$$= -(3)(-2)(5) = 30.$$

13. We have

$$\det \begin{bmatrix} 1 & 2 & 1 \\ 1 & 1 & 2 \\ 3 & 4 & 8 \end{bmatrix} = \det \begin{bmatrix} 1 & 2 & 1 \\ 0 & -1 & 1 \\ 0 & -2 & 5 \end{bmatrix} = \det \begin{bmatrix} 1 & 2 & 1 \\ 0 & -1 & 1 \\ 0 & 0 & 3 \end{bmatrix}$$

$$= 1(-1)(3) = -3.$$

17. We have

$$\det \begin{bmatrix} 0 & 4 & -1 & 1 \\ -3 & 1 & 1 & 2 \\ 1 & 0 & -2 & 3 \\ 2 & 3 & 0 & 1 \end{bmatrix} = - \det \begin{bmatrix} 1 & 0 & -2 & 3 \\ -3 & 1 & 1 & 2 \\ 0 & 4 & -1 & 1 \\ 2 & 3 & 0 & 1 \end{bmatrix} = - \det \begin{bmatrix} 1 & 0 & -2 & 3 \\ 0 & 1 & -5 & 11 \\ 0 & 4 & -1 & 1 \\ 0 & 3 & 4 & -5 \end{bmatrix}$$

$$= - \det \begin{bmatrix} 1 & 0 & -2 & 3 \\ 0 & 1 & -5 & 11 \\ 0 & 0 & 19 & -43 \\ 0 & 0 & 19 & -38 \end{bmatrix} = - \det \begin{bmatrix} 1 & 0 & -2 & 3 \\ 0 & 1 & -5 & 11 \\ 0 & 0 & 19 & -43 \\ 0 & 0 & 0 & 5 \end{bmatrix}$$

$$= -1(1)(19)(5) = -95.$$

21. We have $\det \begin{bmatrix} c & 6 \\ 2 & c+4 \end{bmatrix} = c(c+4) - 12 = c^2 + 4c - 12 = (c+6)(c-2)$.

So the matrix is not invertible if $c = -6$ or $c = 2$.

25. We have

$$\det \begin{bmatrix} 1 & -1 & 2 \\ -1 & 0 & 4 \\ 2 & 1 & c \end{bmatrix} = \begin{bmatrix} 1 & -1 & 2 \\ 0 & -1 & 6 \\ 0 & 3 & c-4 \end{bmatrix} = \begin{bmatrix} 1 & -1 & 2 \\ 0 & -1 & 6 \\ 0 & 0 & c+14 \end{bmatrix}$$

$$= 1(-1)(c+14) = -(c+14).$$

So the matrix is not invertible if $c = -14$.

29. We have

$$x_1 = \frac{\det \begin{bmatrix} 6 & 2 \\ -3 & 4 \end{bmatrix}}{\det \begin{bmatrix} 1 & 2 \\ 3 & 4 \end{bmatrix}} = \frac{6(4) - 2(-3)}{1(4) - 2(3)} = -15$$

and

$$x_2 = \frac{\det \begin{bmatrix} 1 & 6 \\ 3 & -3 \end{bmatrix}}{\det \begin{bmatrix} 1 & 2 \\ 3 & 4 \end{bmatrix}} = \frac{1(-3) - 6(3)}{-2} = 10.5.$$

33. We have

$$x_1 = \frac{\det \begin{bmatrix} 6 & 0 & -2 \\ -5 & 1 & 3 \\ 4 & 2 & 1 \end{bmatrix}}{\det \begin{bmatrix} 1 & 0 & -2 \\ -1 & 1 & 3 \\ 0 & 2 & 1 \end{bmatrix}} = \frac{\det \begin{bmatrix} 6 & 0 & -2 \\ -5 & 1 & 3 \\ 14 & 0 & -5 \end{bmatrix}}{\det \begin{bmatrix} 1 & 0 & -2 \\ 0 & 1 & 1 \\ 0 & 2 & 1 \end{bmatrix}} = \frac{6(-5) - (-2)(14)}{1(1) - 1(2)} = 2,$$

$$x_2 = \frac{\det \begin{bmatrix} 1 & 6 & -2 \\ -1 & -5 & 3 \\ 0 & 4 & 1 \end{bmatrix}}{\det \begin{bmatrix} 1 & 0 & -2 \\ -1 & 1 & 3 \\ 0 & 2 & 1 \end{bmatrix}} = \frac{\det \begin{bmatrix} 1 & 6 & -2 \\ 0 & 1 & 1 \\ 0 & 4 & 1 \end{bmatrix}}{-1} = \frac{1(1) - 1(4)}{-1} = 3,$$

and

$$x_3 = \frac{\det \begin{bmatrix} 1 & 0 & 6 \\ -1 & 1 & -5 \\ 0 & 2 & 4 \end{bmatrix}}{\det \begin{bmatrix} 1 & 0 & -2 \\ -1 & 1 & 3 \\ 0 & 2 & 1 \end{bmatrix}} = \frac{\det \begin{bmatrix} 1 & 0 & 6 \\ 0 & 1 & 1 \\ 0 & 2 & 4 \end{bmatrix}}{-1} = \frac{1(4) - 1(2)}{-1} = -2.$$

37. Take $A = I_2$ and $k = 3$. Then

$$\det kA = \det \begin{bmatrix} 3 & 0 \\ 0 & 3 \end{bmatrix} = 3 \cdot 3 = 9,$$

whereas

$$k \cdot \det A = 3 \cdot 1 = 3.$$

41. By (b) and (d) of Theorem 3.4, we have

$$\begin{aligned} \det(B^{-1}AB) &= (\det B^{-1})(\det A)(\det B) \\ &= (\det A)(\det B^{-1})(\det B) \\ &= (\det A)\left(\frac{1}{\det B}\right)(\det B) \\ &= \det A. \end{aligned}$$

45. We have

$$\det \begin{bmatrix} 1 & a & a^2 \\ 1 & b & b^2 \\ 1 & c & c^2 \end{bmatrix} = \det \begin{bmatrix} 1 & a & a^2 \\ 0 & b-a & b^2-a^2 \\ 0 & c-a & c^2-a^2 \end{bmatrix}$$

$$= (b-a)(c-a) \det \begin{bmatrix} 1 & a & a^2 \\ 0 & 1 & b+a \\ 0 & 1 & c+a \end{bmatrix}$$

$$= (b-a)(c-a) \det \begin{bmatrix} 1 & a & a^2 \\ 0 & 1 & b+a \\ 0 & 0 & c-b \end{bmatrix}$$

$$\begin{aligned} &= (b-a)(c-a) \cdot (1)(1)(c-b) \\ &= (b-a)(c-a)(c-b). \end{aligned}$$

67

49. Let A be an $n \times n$ matrix, and let B be obtained by multiplying each entry of row r of A by the scalar k. Suppose that c_{ij} is the (i, j)-cofactor of A. Because the entries of A and B differ only in row r, the (r, j)-cofactor of B is c_{rj} for $j = 1, 2, \ldots, n$. Evaluating $\det B$ by cofactor expansion along row r gives

$$
\begin{aligned}
\det B &= b_{r1} c_{r1} + b_{r2} c_{r2} + \cdots + b_{rn} c_{rn} \\
&= (k a_{r1}) c_{r1} + (k a_{r2}) c_{r2} + \cdots + (k a_{rn}) c_{rn} \\
&= k(a_{r1} c_{r1} + a_{r2} c_{r2} + \cdots + a_{rn} c_{rn}) \\
&= k(\det A).
\end{aligned}
$$

53. (a) We have

$$
\begin{bmatrix}
0.0 & -3.0 & -2 & -5 \\
2.4 & 3.0 & -6 & 9 \\
-4.8 & 6.3 & 4 & -2 \\
9.6 & 1.5 & 5 & 9
\end{bmatrix}
\longrightarrow
\begin{bmatrix}
2.4 & 3.0 & -6 & 9 \\
0.0 & -3.0 & -2 & -5 \\
-4.8 & 6.3 & 4 & -2 \\
9.6 & 1.5 & 5 & 9
\end{bmatrix}
$$

$$
\longrightarrow
\begin{bmatrix}
2.4 & 3.0 & -6 & 9 \\
0.0 & -3.0 & -2 & -5 \\
0.0 & 12.3 & -8 & 16 \\
0.0 & -10.5 & 29 & -27
\end{bmatrix}
\longrightarrow
\begin{bmatrix}
2.4 & 3 & -6.0 & 9.0 \\
0.0 & -3 & -2.0 & -5.0 \\
0.0 & 0 & -16.2 & -4.5 \\
0.0 & 0 & 36.0 & -9.5
\end{bmatrix}
$$

$$
\longrightarrow
\begin{bmatrix}
2.4 & 3 & -6.0 & 9.0 \\
0.0 & -3 & -2.0 & -5.0 \\
0.0 & 0 & -16.2 & -4.5 \\
0.0 & 0 & 0.0 & -19.5
\end{bmatrix} .
$$

(b) $\det A = (-1)^1 (2.4)(-3)(-16.2)(-19.5) = 2274.48$

CHAPTER 3 REVIEW

1. (a) False, $\det \begin{bmatrix} a & b \\ c & d \end{bmatrix} = ad - bc$.

(b) False, for $n \geq 2$, the (i, j)-cofactor of an $n \times n$ matrix A equals $(-1)^{i+j}$ times the determinant of the $(n-1) \times (n-1)$ matrix obtained by deleting row i and column j from A.

(c) True

(d) False, consider $A = [\mathbf{e}_1 \ \mathbf{0}]$ and $B = [\mathbf{0} \ \mathbf{e}_2]$.

(e) True

(f) False, if B is obtained by interchanging two rows of an $n \times n$ matrix A, then $\det B = -\det A$.

(g) False, an $n \times n$ matrix is invertible if and only if its determinant is nonzero.

(h) True

(i) False, for any invertible matrix A, $\det A^{-1} = \dfrac{1}{\det A}$.

(j) False, for any $n \times n$ matrix A and scalar c, $\det cA = c^n(\det A)$.

5. The $(3,1)$-cofactor of the matrix is

$$(-1)^{3+1} \det \begin{bmatrix} -1 & 2 \\ 2 & -1 \end{bmatrix} = 1[(-1)(-1) - 2(2)] = -3.$$

9. The determinant of the given matrix is

$$1(-1)^{1+1} \det \begin{bmatrix} 2 & -1 \\ 1 & 3 \end{bmatrix} + (-1)(-1)^{1+2} \det \begin{bmatrix} -1 & 2 \\ 1 & 3 \end{bmatrix} + 2(-1)^{1+3} \det \begin{bmatrix} -1 & 2 \\ 2 & -1 \end{bmatrix}$$

$$= 1[2(3) - (-1)(1)] + [(-1)3 - 2(1)] + 2[(-1)(-1) - 2(2)]$$
$$= -4.$$

13. We have

$$\det \begin{bmatrix} 1 & -3 & 1 \\ 4 & -2 & 1 \\ 2 & 5 & -1 \end{bmatrix} = \det \begin{bmatrix} 1 & -3 & 1 \\ 0 & 10 & -3 \\ 0 & 11 & -3 \end{bmatrix}$$

$$= 1(-1)^{1+1} \det \begin{bmatrix} 10 & -3 \\ 11 & -3 \end{bmatrix} = 1[10(-3) - (-3)(11)] = 3.$$

17. We have

$$\begin{bmatrix} c+4 & -1 & c+5 \\ -3 & 3 & -4 \\ c+6 & -3 & c+7 \end{bmatrix} = \begin{bmatrix} c+4 & -1 & c+5 \\ 3c+9 & 0 & 3c+11 \\ -2c-6 & 0 & -2c-8 \end{bmatrix}$$

$$= (-1)(-1)^{1+2} \det \begin{bmatrix} 3c+9 & 3c+11 \\ -2c-6 & -2c-8 \end{bmatrix}$$

$$= -1(c+3) \det \begin{bmatrix} 3 & 3c+11 \\ -2 & -2c-8 \end{bmatrix}$$

$$= (c+3)[3(-2c-8) - (3c+11)(-2)]$$
$$= -2(c+3).$$

So the matrix is not invertible if and only if $c = -3$.

19. The area of the parallelogram in $\mathcal{R}^2$ determined by $\begin{bmatrix} 3 \\ 7 \end{bmatrix}$ and $\begin{bmatrix} 4 \\ 1 \end{bmatrix}$ is

$$\left| \det \begin{bmatrix} 3 & 4 \\ 7 & 1 \end{bmatrix} \right| = |3(1) - 4(7)| = 25.$$

21. We have

$$x_1 = \frac{\det \begin{bmatrix} 5 & 1 \\ -6 & 3 \end{bmatrix}}{\det \begin{bmatrix} 2 & 1 \\ -4 & 3 \end{bmatrix}} = \frac{5(3) - 1(-6)}{2(3) - 1(-4)} = 2.1$$

and

$$x_2 = \frac{\det \begin{bmatrix} 2 & 5 \\ -4 & -6 \end{bmatrix}}{\det \begin{bmatrix} 2 & 1 \\ -4 & 3 \end{bmatrix}} = \frac{2(-6) - 5(-4)}{10} = 0.8.$$

25. If $\det A = 5$, then $\det 2A = 2^3(\det A) = 8(5) = 40$ because we can remove a factor of 2 from each row of $2A$ and apply Theorem 3.3(b).

29. We have

$$\det \begin{bmatrix} a_{11} + 5a_{31} & a_{12} + 5a_{32} & a_{13} + 5a_{33} \\ 4a_{21} & 4a_{22} & 4a_{23} \\ a_{31} - 2a_{21} & a_{32} - 2a_{22} & a_{33} - 2a_{23} \end{bmatrix}$$

$$= 4 \cdot \det \begin{bmatrix} a_{11} + 5a_{31} & a_{12} + 5a_{32} & a_{13} + 5a_{33} \\ a_{21} & a_{22} & a_{23} \\ a_{31} - 2a_{21} & a_{32} - 2a_{22} & a_{33} - 2a_{22} \end{bmatrix}$$

$$= 4 \cdot \det \begin{bmatrix} a_{11} + 5a_{31} & a_{12} + 5a_{32} & a_{13} + 5a_{33} \\ a_{21} & a_{22} & a_{23} \\ a_{31} & a_{32} & a_{33} \end{bmatrix}$$

$$= 4 \cdot \det \begin{bmatrix} a_{11} & a_{12} & a_{13} \\ a_{21} & a_{22} & a_{23} \\ a_{31} & a_{32} & a_{33} \end{bmatrix} = 4(\det A) = 4(5) = 20.$$

33. We have

$$\det \begin{bmatrix} 1 & x & y \\ 1 & x_1 & y_1 \\ 0 & 1 & m \end{bmatrix} = \det \begin{bmatrix} 1 & x & y \\ 0 & x_1 - x & y_1 - y \\ 0 & 1 & m \end{bmatrix}$$

$$= 1(-1)^{1+1} \det \begin{bmatrix} x_1 - x & y_1 - y \\ 1 & m \end{bmatrix}$$

$$= (x_1 - x)m - (y_1 - y)(1)$$
$$= m(x_1 - x) - (y_1 - y).$$

So the given equation equals

$$m(x_1 - x) - (y_1 - y) = 0$$
$$-m(x - x_1) + (y - y_1) = 0$$
$$y - y_1 = m(x - x_1),$$

which is the equation of the line through (x_1, y_1) with slope m.

Chapter 4

Subspaces and Their Properties

4.1 SUBSPACES

1. (a) True
   (b) False, the empty set does not contain **0**.
   (c) False, $\{0\}$ is called the *zero subspace*.
   (d) True
   (e) True
   (f) False, the column space of an $m \times n$ matrix is contained in $\mathcal{R}^m$.
   (g) False, the row space of an $m \times n$ matrix is contained in $\mathcal{R}^n$.
   (h) False, the column space of an $m \times n$ matrix equals $\{A\mathbf{v} : \mathbf{v} \text{ is in } \mathcal{R}^n\}$.
   (i) True
   (j) True
   (k) False, the range of a linear transformation is always a subspace.

5. Since

$$
\begin{bmatrix} -s + t \\ 2s - t \\ s + 3t \end{bmatrix} = s \begin{bmatrix} -1 \\ 2 \\ 1 \end{bmatrix} + t \begin{bmatrix} 1 \\ -1 \\ 3 \end{bmatrix},
$$

a spanning set for the subspace is

$$
\left\{ \begin{bmatrix} -1 \\ 2 \\ 1 \end{bmatrix}, \begin{bmatrix} 1 \\ -1 \\ 3 \end{bmatrix} \right\}.
$$

9. Since

$$
\begin{bmatrix} 2s - 5t \\ 3r + s - 2t \\ r - 4s + 3t \\ -r + 2s \end{bmatrix} = r \begin{bmatrix} 0 \\ 3 \\ 1 \\ -1 \end{bmatrix} + s \begin{bmatrix} 2 \\ 1 \\ -4 \\ 2 \end{bmatrix} + t \begin{bmatrix} -5 \\ -2 \\ 3 \\ 0 \end{bmatrix},
$$

a spanning set for the subspace is

$$\left\{ \begin{bmatrix} 0 \\ 3 \\ 1 \\ -1 \end{bmatrix}, \begin{bmatrix} 2 \\ 1 \\ -4 \\ 2 \end{bmatrix}, \begin{bmatrix} -5 \\ -2 \\ 3 \\ 0 \end{bmatrix} \right\}.$$

13. If $\mathbf{0}$ were in the set, then therewould be scalars s and t such that

$$\begin{aligned} 3s \quad\quad &= 2 \\ 2s + 4t &= 0 \\ -t &= 0. \end{aligned}$$

Since the system has no solutions, $\mathbf{0}$ is not in the given set.

17. Consider

$$\mathbf{v} = \begin{bmatrix} 2 \\ 1 \\ 2 \end{bmatrix} \quad \text{and} \quad 3\mathbf{v} = \begin{bmatrix} 6 \\ 3 \\ 6 \end{bmatrix}.$$

The vector $\mathbf{v}$ belongs to the given set because $2 = 1(2)$. However, $3\mathbf{v}$ does not belong to the given set because $6 \neq 3(6)$.

21. For the given vector $\mathbf{v}$, we have $A\mathbf{v} \neq \mathbf{0}$. Hence $\mathbf{v}$ does not belong to Null A.

25. Vector $\mathbf{b} = \begin{bmatrix} 1 \\ -4 \\ 2 \end{bmatrix}$ belongs to Col A if and only if $A\mathbf{x} = \mathbf{b}$ is consistent. Since $\begin{bmatrix} -7 \\ -4 \\ 0 \\ 0 \end{bmatrix}$ is

a solution to this system, $\mathbf{b}$ belongs to Col A.

29. The reduced row echelon form of the given matrix A is

$$\begin{bmatrix} 1 & 0 & 1 & 0 & -1 \\ 0 & 1 & -1 & 0 & 0 \\ 0 & 0 & 0 & 1 & 1 \end{bmatrix}.$$

Since the parametric representation of the general solution to $A\mathbf{x} = \mathbf{0}$ is

$$\begin{bmatrix} x_1 \\ x_2 \\ x_3 \\ x_4 \\ x_5 \end{bmatrix} = x_3 \begin{bmatrix} -1 \\ 1 \\ 1 \\ 0 \\ 0 \end{bmatrix} + x_5 \begin{bmatrix} 1 \\ 0 \\ 0 \\ -1 \\ 1 \end{bmatrix},$$

a spanning set for Null A is

$$\left\{ \begin{bmatrix} -1 \\ 1 \\ 1 \\ 0 \\ 0 \end{bmatrix}, \begin{bmatrix} 1 \\ 0 \\ 0 \\ -1 \\ 1 \end{bmatrix} \right\}.$$

33. From the matrix R in the solution to Exercise 29, we see that the pivot columns of the given matrix are columns 1, 2, and 4. Choosing each of these columns and exactly one of the other columns gives a spanning set for the column space of the matrix that contains exactly four vectors. (See Theorems 2.5(b) and 1.6(c).) One such spanning set is the set containing the first four columns of the given matrix.

37. Consider $A = \begin{bmatrix} 1 & 1 \\ 2 & 2 \end{bmatrix}$. The reduced row echelon form of A is $R = \begin{bmatrix} 1 & 1 \\ 0 & 0 \end{bmatrix}$. Clearly $\begin{bmatrix} 1 \\ 2 \end{bmatrix}$ belongs to Col A but not to Col R; so Col $A \neq$ Col R.

41. Let V and W be subspaces of $\mathcal{R}^n$. Since $\mathbf{0}$ is contained in both V and W, $\mathbf{0}$ is contained in $V \cap W$. Let $\mathbf{v}$ and $\mathbf{w}$ be contained in $V \cap W$. Then $\mathbf{v}$ and $\mathbf{w}$ are contained in both V and W, and so $\mathbf{v} + \mathbf{w}$ is contained in both V and W. Thus $\mathbf{v} + \mathbf{w}$ is contained in $V \cap W$. Finally, for any scalar c, $c\mathbf{v}$ is contained in both V and W; so $c\mathbf{v}$ is in $V \cap W$. It follows that $V \cap W$ is a subspace of $\mathcal{R}^n$.

45. The standard matrix of T is

$$A = \begin{bmatrix} 1 & 2 & -1 \end{bmatrix}.$$

The range of T is spanned by the columns of A; so $\{1, 2, -1\}$ is a spanning set for the range of T.

Note that A is in reduced row echelon form. The general solution of $A\mathbf{x} = \mathbf{0}$ is

$$\begin{array}{ll} x_1 = -2x_2 + x_3 \\ x_2 & \text{free} \\ x_3 & \text{free,} \end{array}$$

and its parametric representation is

$$\begin{bmatrix} x_1 \\ x_2 \\ x_3 \end{bmatrix} = x_2 \begin{bmatrix} -2 \\ 1 \\ 0 \end{bmatrix} + x_3 \begin{bmatrix} 1 \\ 0 \\ 1 \end{bmatrix}.$$

Hence

$$\left\{ \begin{bmatrix} -2 \\ 1 \\ 0 \end{bmatrix}, \begin{bmatrix} 1 \\ 0 \\ 1 \end{bmatrix} \right\}$$

is a spanning set for the null spaces of both A and T.

49. The standard matrix of T is

$$A = \begin{bmatrix} 1 & 1 & -1 \\ 0 & 0 & 0 \\ 2 & 0 & -1 \end{bmatrix}.$$

Since the range of T is spanned by the columns of A, a spanning set for the range of T is

$$\left\{ \begin{bmatrix} 1 \\ 0 \\ 2 \end{bmatrix}, \begin{bmatrix} 1 \\ 0 \\ 0 \end{bmatrix}, \begin{bmatrix} -1 \\ 0 \\ -1 \end{bmatrix} \right\}.$$

Since the reduced row echelon form of A is

$$\begin{bmatrix} 1 & 0 & -.5 \\ 0 & 1 & -.5 \\ 0 & 0 & 0 \end{bmatrix},$$

the parametric representation of the genereal solution to $A\mathbf{x} = \mathbf{0}$ is

$$\begin{bmatrix} x_1 \\ x_2 \\ x_3 \end{bmatrix} = x_2 \begin{bmatrix} .5 \\ .5 \\ 1 \end{bmatrix}.$$

Hence a spanning set for Null A is

$$\left\{ \begin{bmatrix} .5 \\ .5 \\ 1 \end{bmatrix} \right\} \quad \text{or} \quad \left\{ \begin{bmatrix} 1 \\ 1 \\ 2 \end{bmatrix} \right\}.$$

53. Denote the given set by V. Since $2u_1 + 5u_2 - 4u_3 = 0$ for $u_1 = u_2 = u_3 = 0$, we see that $\mathbf{0}$ belongs to V. Let $\mathbf{u}$ and $\mathbf{v}$ belong to V. Then $2u_1 + 5u_2 - 4u_3 = 0$ and $2v_1 + 5v_2 - 4v_3 = 0$. Now

$$\mathbf{u} + \mathbf{v} = \begin{bmatrix} u_1 + v_1 \\ u_2 + v_2 \\ u_3 + v_3 \end{bmatrix},$$

and

$$2(u_1 + v_1) + 5(u_2 + v_2) - 4(u_3 + v_3) = (2u_1 + 5u_2 - 4u_3) + (2v_1 + 5v_2 - 4v_3)$$
$$= 0 + 0$$
$$= 0.$$

So $\mathbf{u} + \mathbf{v}$ belongs to V. Thus V is closed under vector addition.

For any scalar c,

$$cu = \begin{bmatrix} cu_1 \\ cu_2 \\ cu_3 \end{bmatrix},$$

and

$$2(cu_1) + 5(cu_2) - 4(cu_3) = c(2u_1 + 5u_2 - 4u_3)$$
$$= c(0)$$
$$= 0.$$

Thus cu belongs to V, and hence V is also closed under scalar multiplication.

Since V is a subset of $\mathcal{R}^3$ that contains $\mathbf{0}$ and is closed under both vector addition and scalar multiplication, V is a subspace of $\mathcal{R}^3$.

57. Let V denote the null space of T. Since $T(\mathbf{0}) = \mathbf{0}$ by Theorem 2.9(a), $\mathbf{0}$ is in V. If $\mathbf{u}$ and $\mathbf{v}$ are in V, then $T(\mathbf{u}) = T(\mathbf{v}) = \mathbf{0}$. Hence

$$T(\mathbf{u} + \mathbf{v}) = T(\mathbf{u}) + T(\mathbf{v}) = \mathbf{0} + \mathbf{0} = \mathbf{0};$$

so $\mathbf{u} + \mathbf{v}$ is in V. Finally, for any scalar c and any vector $\mathbf{u}$ in V, we have

$$T(c\mathbf{u}) = cT(\mathbf{u}) = c(\mathbf{0}) = \mathbf{0};$$

so $c\mathbf{u}$ is in V. Thus V is also closed under scalar multiplication. Since V is a subset of $\mathcal{R}^n$ that contains $\mathbf{0}$ and is closed under both vector addition and scalar multiplication, V is a subspace of $\mathcal{R}^n$.

61. (a) The system $A\mathbf{x} = \mathbf{u}$ is consistent since the reduced row echelon form of $[A \ \mathbf{u}]$ contains no row whose only nonzero entry lies in the last column. Hence $\mathbf{u}$ belongs to $\text{Col} \, A$.

   (b) On the other hand, $A\mathbf{x} = \mathbf{v}$ is not consistent, and so $\mathbf{v}$ does not belong to $\text{Col} \, A$.

4.2   BASIS AND DIMENSION

1. (a) False, every nonzero subspace of $\mathcal{R}^n$ has infinitely many bases.
   (b) True
   (c) False, a basis for a subspace is a spanning set that is as *small* as possible.
   (d) False, for $\mathcal{S}$ to be a basis for V, $\mathcal{S}$ must be a subset of V.
   (e) True
   (f) True
   (g) True
   (h) False, the *pivot columns* of any matrix form a basis for its column space.

(i) False, if $A = \begin{bmatrix} 1 & 2 \\ 1 & 2 \end{bmatrix}$, then the pivot columns of the reduced row echelon form of A do not form a basis for Col A.

(j) True

3. A set of 4 vectors from $\mathcal{R}^3$ must be linearly dependent by Theorem 1.9′ on page 213.

5. Since

$$\begin{bmatrix} s \\ -2s \end{bmatrix} = s \begin{bmatrix} 1 \\ -2 \end{bmatrix}$$

and $\left\{ \begin{bmatrix} 1 \\ -2 \end{bmatrix} \right\}$ is linearly independent, this set is a basis for the given subspace.

9. Let

$$A = \begin{bmatrix} 1 & 2 & 1 \\ 2 & 1 & -4 \\ 1 & 3 & 3 \end{bmatrix}.$$

Then the given subspace is Col A, and so a basis for the given subspace can be obtained by choosing the pivot columns of A. Since the reduced row echelon form of A is

$$\begin{bmatrix} 1 & 0 & -3 \\ 0 & 1 & 2 \\ 0 & 0 & 0 \end{bmatrix},$$

this basis is

$$\left\{ \begin{bmatrix} 1 \\ 2 \\ 1 \end{bmatrix}, \begin{bmatrix} 2 \\ 1 \\ 3 \end{bmatrix} \right\}.$$

13. As in Exercise 9, form a 4 × 5 matrix whose columns are the vectors in the given set. The pivot columns of this matrix form a basis for the given subspace. Since the reduced row echelon form of this matrix is

$$\begin{bmatrix} 1 & 0 & -5 & 2 & 0 \\ 0 & 1 & 3 & -1 & 0 \\ 0 & 0 & 0 & 0 & 1 \\ 0 & 0 & 0 & 0 & 0 \end{bmatrix},$$

one basis for the given subspace is

$$\left\{ \begin{bmatrix} 1 \\ 0 \\ -1 \\ 2 \end{bmatrix}, \begin{bmatrix} 1 \\ 1 \\ -2 \\ 1 \end{bmatrix}, \begin{bmatrix} 0 \\ 1 \\ -1 \\ 2 \end{bmatrix} \right\}.$$

Chapter 4 Subspaces and Their Properties

17. The reduced row echelon form of the given matrix A is

$$\begin{bmatrix} 1 & 0 & 0 & 4 \\ 0 & 1 & 0 & 4 \\ 0 & 0 & 1 & 1 \\ 0 & 0 & 0 & 0 \end{bmatrix}.$$

(a) Hence the first three columns of the given matrix are its pivot columns, and so

$$\left\{ \begin{bmatrix} -1 \\ 2 \\ 1 \\ 0 \end{bmatrix}, \begin{bmatrix} 1 \\ 0 \\ -1 \\ 1 \end{bmatrix}, \begin{bmatrix} 2 \\ -5 \\ -1 \\ -2 \end{bmatrix} \right\}$$

is a basis for the column space of A.

(b) The null space of A is the solution set of $A\mathbf{x} = \mathbf{0}$. Since the parametric representation of the general solution of $A\mathbf{x} = \mathbf{0}$ is

$$\begin{bmatrix} x_1 \\ x_2 \\ x_3 \\ x_4 \end{bmatrix} = x_4 \begin{bmatrix} -4 \\ -4 \\ -1 \\ 1 \end{bmatrix},$$

the set

$$\left\{ \begin{bmatrix} -4 \\ -4 \\ -1 \\ 1 \end{bmatrix} \right\}$$

is a basis for the null space of A.

21. The standard matrix of T is

$$A = \begin{bmatrix} 1 & -2 & 1 & 1 \\ 2 & -5 & 1 & 3 \\ 1 & -3 & 0 & 2 \end{bmatrix}.$$

(a) The range of T equals the column space of A; so we proceed as in Exercise 17. The reduced row echelon form of A is

$$R = \begin{bmatrix} 1 & 0 & 3 & -1 \\ 0 & 1 & 1 & -1 \\ 0 & 0 & 0 & 0 \end{bmatrix}.$$

Hence the set of pivot columns of A,

$$\left\{ \begin{bmatrix} 1 \\ 2 \\ 1 \end{bmatrix}, \begin{bmatrix} -2 \\ -5 \\ -3 \end{bmatrix} \right\},$$

is a basis for the range of T.

78

(b) Since the null space of T is the same as the null space of A, we must determine the parametric representation of the general solution of $A\mathbf{x} = \mathbf{0}$. This representation is:

$$\begin{bmatrix} x_1 \\ x_2 \\ x_3 \\ x_4 \end{bmatrix} = x_3 \begin{bmatrix} -3 \\ -1 \\ 1 \\ 0 \end{bmatrix} + x_4 \begin{bmatrix} 1 \\ 1 \\ 0 \\ 1 \end{bmatrix}.$$

Hence

$$\left\{ \begin{bmatrix} -3 \\ -1 \\ 1 \\ 0 \end{bmatrix}, \begin{bmatrix} 1 \\ 1 \\ 0 \\ 1 \end{bmatrix} \right\}$$

is a basis for the null space of T.

25. The standard matrix of T is

$$A = \begin{bmatrix} 1 & 2 & 3 & 0 & 4 \\ 3 & 1 & -1 & 0 & -3 \\ 7 & 4 & 1 & 0 & -2 \end{bmatrix},$$

and the reduced row echelon form of A is

$$\begin{bmatrix} 1 & 0 & -1 & 0 & -2 \\ 0 & 1 & 2 & 0 & 3 \\ 0 & 0 & 0 & 0 & 0 \end{bmatrix}.$$

(a) As in Exercise 21, the set of pivot columns of A,

$$\left\{ \begin{bmatrix} 1 \\ 3 \\ 7 \end{bmatrix}, \begin{bmatrix} 2 \\ 1 \\ 4 \end{bmatrix} \right\},$$

is a basis for the range of T.

(b) The parametric representation of the general solution of $A\mathbf{x} = \mathbf{0}$ is

$$\begin{bmatrix} x_1 \\ x_2 \\ x_3 \\ x_4 \\ x_5 \end{bmatrix} = x_3 \begin{bmatrix} 1 \\ -2 \\ 1 \\ 0 \\ 0 \end{bmatrix} + x_4 \begin{bmatrix} 0 \\ 0 \\ 0 \\ 1 \\ 0 \end{bmatrix} + x_5 \begin{bmatrix} 2 \\ -3 \\ 0 \\ 0 \\ 1 \end{bmatrix}.$$

Thus

$$\left\{ \begin{bmatrix} 1 \\ -2 \\ 1 \\ 0 \\ 0 \end{bmatrix}, \begin{bmatrix} 0 \\ 0 \\ 0 \\ 1 \\ 0 \end{bmatrix}, \begin{bmatrix} 2 \\ -3 \\ 0 \\ 0 \\ 1 \end{bmatrix} \right\}$$

is a basis for the null space of T.

29. Let A denote the matrix in Exercise 15,

$$\mathbf{u} = \begin{bmatrix} 0 \\ 1 \\ 1 \\ 1 \end{bmatrix}, \quad \text{and} \quad \mathbf{v} = \begin{bmatrix} 2 \\ 2 \\ 1 \\ 1 \end{bmatrix}.$$

Since $A\mathbf{u} = \mathbf{0}$ and $A\mathbf{v} = \mathbf{0}$, $\{\mathbf{u}, \mathbf{v}\}$ is a subset of Null A. Moreover, neither $\mathbf{u}$ nor $\mathbf{v}$ is a multiple of the other; so $\{\mathbf{u}, \mathbf{v}\}$ is linearly independent. From Exercise 15, we see that the dimension of Null A is 2. Since $\{\mathbf{u}, \mathbf{v}\}$ is a linearly independent subset of Null A that contains two vectors, it must be a basis for Null A by Theorem 4.5.

33. Let V denote the given subspace of $\mathcal{R}^n$. Clearly $\mathcal{B} = \{\mathbf{e}_3, \mathbf{e}_4, \ldots, \mathbf{e}_n\}$ is a subset of V, and $\mathcal{B}$ is linearly independent because every column of $[\mathbf{e}_3 \ \mathbf{e}_4 \ \cdots \ \mathbf{e}_n]$ is a pivot column. Moreover, $\mathcal{B}$ spans V, for if $\mathbf{v}$ is in V, then

$$\mathbf{v} = \begin{bmatrix} 0 \\ 0 \\ v_3 \\ \vdots \\ v_n \end{bmatrix} = v_3 \mathbf{e}_3 + \cdots + v_n \mathbf{e}_n.$$

Since $\mathcal{B}$ is a linearly independent spanning set for V, we see that $\mathcal{B}$ is a basis for V. Hence the dimension of V equals the number of vectors in $\mathcal{B}$, which is $n - 2$.

37. Clearly $\mathbf{v}$ belongs to Span $\mathcal{A} = V$. Thus $\mathcal{B} = \{\mathbf{v}, \mathbf{u}_2, \mathbf{u}_3, \ldots, \mathbf{u}_k\}$ is a subset of V, because $\mathbf{u}_2, \mathbf{u}_3, \ldots, \mathbf{u}_k$ belong to $\mathcal{A}$, which is a subset of V.

We claim that $\mathcal{B}$ is linearly independent. Suppose that $c_1, c_2, \ldots, c_k$ are scalars such that

$$c_1 \mathbf{v} + c_2 \mathbf{u}_2 + \cdots + c_k \mathbf{u}_k = \mathbf{0}.$$

Then

$$c_1(\mathbf{u}_1 + \mathbf{u}_2 + \cdots + \mathbf{u}_k) + c_2 \mathbf{u}_2 + \cdots + c_k \mathbf{u}_k = \mathbf{0}$$
$$c_1 \mathbf{u}_1 + (c_1 + c_2)\mathbf{u}_2 + \cdots + (c_1 + c_k)\mathbf{u}_k = \mathbf{0}.$$

Since $\mathcal{A}$ is linearly independent, it follows that $c_1 = 0$, $c_1 + c_2 = 0$, $\cdots$, $c_1 + c_k = 0$. Hence $c_1 = c_2 = \cdots = c_k = 0$, proving that $\mathcal{B}$ is linearly independent. Since $\mathcal{B}$ contains k vectors, it follows from Theorem 4.5 that $\mathcal{B}$ is a basis for V.

41. The zero subspace $\{\mathbf{0}\}$ has no basis because its only spanning set $\{\mathbf{0}\}$ is linearly dependent.

45. Let

$$A = \begin{bmatrix} 1 & -1 & 2 & 1 \\ 2 & -2 & 4 & 2 \\ -3 & 3 & -6 & -3 \end{bmatrix}.$$

Since the reduced row echelon form of A is

$$\begin{bmatrix} 1 & -1 & 2 & 1 \\ 0 & 0 & 0 & 0 \\ 0 & 0 & 0 & 0 \end{bmatrix},$$

the parametric representation of the general solution of $A\mathbf{x} = \mathbf{0}$ is

$$\begin{bmatrix} x_1 \\ x_2 \\ x_3 \\ x_4 \end{bmatrix} = x_2 \begin{bmatrix} 1 \\ 1 \\ 0 \\ 0 \end{bmatrix} + x_3 \begin{bmatrix} -2 \\ 0 \\ 1 \\ 0 \end{bmatrix} + x_4 \begin{bmatrix} -1 \\ 0 \\ 0 \\ 1 \end{bmatrix}.$$

Hence

$$\left\{ \begin{bmatrix} 1 \\ 1 \\ 0 \\ 0 \end{bmatrix}, \begin{bmatrix} -2 \\ 0 \\ 1 \\ 0 \end{bmatrix}, \begin{bmatrix} -1 \\ 0 \\ 0 \\ 1 \end{bmatrix} \right\}$$

is a basis for Null A.

Since the reduced row echelon form of

$$\begin{bmatrix} 0 & 1 & -2 & 1 \\ 2 & 1 & 0 & 0 \\ 1 & 0 & 1 & 0 \\ 0 & 0 & 0 & 1 \end{bmatrix}$$

is

$$\begin{bmatrix} 1 & 0 & 1 & 0 \\ 0 & 1 & -2 & 0 \\ 0 & 0 & 0 & 0 \\ 0 & 0 & 0 & 0 \end{bmatrix},$$

it follows from Exercise 42 that

$$\left\{ \begin{bmatrix} 0 \\ 2 \\ 1 \\ 0 \end{bmatrix}, \begin{bmatrix} 1 \\ 1 \\ 0 \\ 0 \end{bmatrix}, \begin{bmatrix} -1 \\ 0 \\ 0 \\ 1 \end{bmatrix} \right\}$$

is a basis for Null A that contains $\mathcal{L}$.

49. The reduced row echelon form of A is

$$\begin{bmatrix} 1 & 0 & -1.2 & 0 & 1.4 \\ 0 & 1 & 2.3 & 0 & -2.9 \\ 0 & 0 & 0.0 & 1 & 0.7 \end{bmatrix}.$$

(a) As in Exercise 17,

$$\left\{ \begin{bmatrix} 0.1 \\ 0.7 \\ -0.5 \end{bmatrix}, \begin{bmatrix} 0.2 \\ 0.9 \\ 0.5 \end{bmatrix}, \begin{bmatrix} 0.5 \\ -0.5 \\ -0.5 \end{bmatrix} \right\}$$

is a basis for the column space of A.

(b) The parametric representation of the general solution of $Ax = 0$ is

$$\begin{bmatrix} x_1 \\ x_2 \\ x_3 \\ x_4 \\ x_5 \end{bmatrix} = x_3 \begin{bmatrix} 1.2 \\ -2.3 \\ 1.0 \\ 0.0 \\ 0.0 \end{bmatrix} + x_5 \begin{bmatrix} -1.4 \\ 2.9 \\ 0.0 \\ -0.7 \\ 1.0 \end{bmatrix}.$$

Hence

$$\left\{ \begin{bmatrix} 1.2 \\ -2.3 \\ 1.0 \\ 0.0 \\ 0.0 \end{bmatrix}, \begin{bmatrix} -1.4 \\ 2.9 \\ 0.0 \\ -0.7 \\ 1.0 \end{bmatrix} \right\}$$

is a basis for Null A.

4.3   THE DIMENSION OF SUBSPACES ASSOCIATED WITH A MATRIX

1. (a) False, the dimensions of the subspaces $V = \text{Span}\{e_1\}$ and $W = \text{Span}\{e_2\}$ of $\mathcal{R}^2$ are both 1, but $V \neq W$.

(b) True

(c) True

(d) False, the dimension of the null space of a matrix equals the nullity of the matrix.

(e) False, the dimension of the column space of a matrix equals the rank of the matrix.

(f) True

(g) True

(h) False, consider $A = \begin{bmatrix} 1 & 2 \\ 1 & 2 \end{bmatrix}$ and the reduced row echelon form of A, which is $\begin{bmatrix} 1 & 2 \\ 0 & 0 \end{bmatrix}$.

(i) True

3. The vectors in the given subspace have the form

$$\begin{bmatrix} -3s + 4t \\ s - 2t \\ 2s \end{bmatrix} = s \begin{bmatrix} -3 \\ 1 \\ 2 \end{bmatrix} + t \begin{bmatrix} 4 \\ -2 \\ 0 \end{bmatrix}.$$

So the given subspace is spanned by

$$\mathcal{B} = \left\{ \begin{bmatrix} -3 \\ 1 \\ 2 \end{bmatrix}, \begin{bmatrix} 4 \\ -2 \\ 0 \end{bmatrix} \right\}.$$

Since neither vector in $\mathcal{B}$ is a multiple of the other, $\mathcal{B}$ is also linearly independent. Therefore $\mathcal{B}$ is a basis for the given subspace.

5. (a) The dimension of Col A equals rank A, which is 2.
    (b) The dimension of Null A equals the nullity of A, which is $4 - 2 = 2$.
    (c) The dimension of Row A equals rank A, which is 2.
    (d) The dimension of Null A^T equals the nullity of A^T. Since A^T is a 4×3 matrix, the nullity of A^T equals

$$3 - \text{rank } A^T = 3 - \text{rank } A = 3 - 2 = 1.$$

9. Clearly rank $A = 1$. So, as in Exercise 5, the answers are:
    (a) 1   (b) 3   (c) 1   (d) 0.

13. The reduced row echelon form of A is

$$\begin{bmatrix} 1 & 0 & 6 & 0 \\ 0 & 1 & -4 & 1 \\ 0 & 0 & 0 & 0 \end{bmatrix}.$$

Hence rank $A = 2$. As in Exercise 5, the answers are:
(a) 2   (b) 2   (c) 2   (d) 1.

17. The standard matrix of T is $A = \begin{bmatrix} 1 & 2 \\ 2 & 1 \end{bmatrix}$, and its reduced row echelon form is I_2.

    (a) Since the range of T equals the column space of A, the dimension of the range of T equals the rank of A, which is 2. Thus, by Theorem 2.12, T is onto.
    (b) The null space of T equals the null space of A. Hence the dimension of the null space of T equals the nullity of A, which is 0. Thus T is one-to-one by Theorem 2.13.

21. The standard matrix of T is

$$\begin{bmatrix} 1 & 0 \\ 2 & 1 \\ 0 & -1 \end{bmatrix},$$

and its reduced row echelon form is

$$\begin{bmatrix} 1 & 0 \\ 0 & 1 \\ 0 & 0 \end{bmatrix}.$$

(a) As in Exercise 17, the dimension of the range of T is 2. Since the codomain of T is $\mathcal{R}^3$, T is not onto.

(b) As in Exercise 17, the dimension of the null space of T is 0. Hence T is one-to-one.

25. Taking $s = \frac{3}{7}$ and $t = -\frac{1}{7}$, we have

$$\begin{bmatrix} 2s - t \\ s + 3t \end{bmatrix} = \begin{bmatrix} 1 \\ 0 \end{bmatrix};$$

and taking $s = \frac{1}{7}$ and $t = \frac{2}{7}$, we have

$$\begin{bmatrix} 2s - t \\ s + 3t \end{bmatrix} = \begin{bmatrix} 0 \\ 1 \end{bmatrix}.$$

Hence $\mathcal{B}$ is contained in V. Moreover, $\mathcal{B}$ is linearly independent. Since the vectors in V have the form

$$\begin{bmatrix} 2s - t \\ s + 3t \end{bmatrix} = s \begin{bmatrix} 2 \\ 1 \end{bmatrix} + t \begin{bmatrix} -1 \\ 3 \end{bmatrix},$$

the set

$$\left\{ \begin{bmatrix} 2 \\ 1 \end{bmatrix}, \begin{bmatrix} -1 \\ 3 \end{bmatrix} \right\}$$

is a basis for V. Hence $\dim V = 2$. Therefore $\mathcal{B}$ is a basis for V because (i), (ii), and (iii) on page 216 are satisfied.

29. Taking $r = 2$, $s = 1$, and $t = 1$, we have

$$\begin{bmatrix} -r + 3s \\ 0 \\ s - t \\ r - 2t \end{bmatrix} = \begin{bmatrix} 1 \\ 0 \\ 0 \\ 0 \end{bmatrix};$$

taking $r = 5$, $s = 2$, and $t = 3$, we have

$$\begin{bmatrix} -r + 3s \\ 0 \\ s - t \\ r - 2t \end{bmatrix} = \begin{bmatrix} 1 \\ 0 \\ -1 \\ -1 \end{bmatrix} ;$$

and taking $r = 2$, $s = 1$, and $t = 0$, we have

$$\begin{bmatrix} -r + 3s \\ 0 \\ s - t \\ r - 2t \end{bmatrix} = \begin{bmatrix} 1 \\ 0 \\ 1 \\ 2 \end{bmatrix} .$$

Hence $\mathcal{B}$ is contained in V.

Since the reduced row echelon form of

$$\begin{bmatrix} 1 & 1 & 1 \\ 0 & 0 & 0 \\ 0 & -1 & 1 \\ 0 & -1 & 2 \end{bmatrix}$$

is

$$\begin{bmatrix} 1 & 0 & 0 \\ 0 & 1 & 0 \\ 0 & 0 & 1 \\ 0 & 0 & 0 \end{bmatrix} ,$$

$\mathcal{B}$ is linearly independent. The vectors in V have the form

$$\begin{bmatrix} -r + 3s \\ 0 \\ s - t \\ r - 2t \end{bmatrix} = r \begin{bmatrix} -1 \\ 0 \\ 0 \\ 1 \end{bmatrix} + s \begin{bmatrix} 3 \\ 0 \\ 1 \\ 0 \end{bmatrix} + t \begin{bmatrix} 0 \\ 0 \\ -1 \\ -2 \end{bmatrix} .$$

It is easily checked that the set

$$\left\{ \begin{bmatrix} -1 \\ 0 \\ 0 \\ 1 \end{bmatrix} , \begin{bmatrix} 3 \\ 0 \\ 1 \\ 0 \end{bmatrix} , \begin{bmatrix} 0 \\ 0 \\ -1 \\ -2 \end{bmatrix} \right\}$$

is linearly independent, and so it is a basis for V. Hence $\dim V = 3$. It follows, as in Exercise 25, that $\mathcal{B}$ is a basis for V.

33. (a) Refer to the solution to Exercise 13. By Theorem 4.6,

$$\left\{\begin{bmatrix} 1 & 0 & 6 & 0 \end{bmatrix}, \begin{bmatrix} 0 & 1 & -4 & 1 \end{bmatrix}\right\}$$

is a basis for Row A. Also, the parametric representation of the general solution to $A\mathbf{x} = \mathbf{0}$ is

$$\begin{bmatrix} x_1 \\ x_2 \\ x_3 \\ x_4 \end{bmatrix} = x_3 \begin{bmatrix} -6 \\ 4 \\ 1 \\ 0 \end{bmatrix} + x_4 \begin{bmatrix} 0 \\ 1 \\ 0 \\ 1 \end{bmatrix}.$$

Thus

$$\left\{\begin{bmatrix} -6 \\ 4 \\ 1 \\ 0 \end{bmatrix}, \begin{bmatrix} 0 \\ 1 \\ 0 \\ 1 \end{bmatrix}\right\}$$

is a basis for Null A.

(b) It is easily checked that set

$$\left\{\begin{bmatrix} 1 \\ 0 \\ 6 \\ 0 \end{bmatrix}, \begin{bmatrix} 0 \\ 1 \\ -4 \\ 1 \end{bmatrix}, \begin{bmatrix} -6 \\ 4 \\ 1 \\ 0 \end{bmatrix}, \begin{bmatrix} 0 \\ 1 \\ 0 \\ 1 \end{bmatrix}\right\}$$

is linearly independent. Since it contains 4 vectors, it is a basis for $\mathcal{R}^4$ by Theorem 4.5.

37. Let $\mathbf{v}$ be in the column space of AB. Then $\mathbf{v} = (AB)\mathbf{u}$ for some $\mathbf{u}$ in $\mathcal{R}^p$. Consider $\mathbf{w} = B\mathbf{u}$. Since $A\mathbf{w} = A(B\mathbf{u}) = (AB)\mathbf{u} = \mathbf{v}$, $\mathbf{v}$ is in the column space of A.

41. Since the ranks of a matrix and its transpose are equal, we have

$$\operatorname{rank} AB = \operatorname{rank} (AB)^T = \operatorname{rank} B^T A^T.$$

By Exercise 39,

$$\operatorname{rank} B^T A^T \leq \operatorname{rank} B^T = \operatorname{rank} B.$$

Combining the preceding results yields $\operatorname{rank} AB \leq \operatorname{rank} B$.

45. (a) Let $\mathbf{v}$ and $\mathbf{w}$ be in $\mathcal{R}^k$. Then

$$T(\mathbf{v} + \mathbf{w}) = T\left(\begin{bmatrix} v_1 + w_1 \\ v_2 + w_2 \\ \vdots \\ v_k + w_k \end{bmatrix}\right)$$

$$= (v_1 + w_1)\mathbf{u}_1 + (v_2 + w_2)\mathbf{u}_2 + \cdots + (v_k + w_k)\mathbf{u}_k$$
$$= (v_1\mathbf{u}_1 + v_2\mathbf{u}_2 + \cdots + v_k\mathbf{u}_k) + (w_1\mathbf{u}_1 + w_2\mathbf{u}_2 + \cdots + w_k\mathbf{u}_k)$$
$$= T(\mathbf{v}) + T(\mathbf{u}).$$

Also, for any scalar c,

$$T(c\mathbf{v}) = T\left(\begin{bmatrix} cv_1 \\ cv_2 \\ \vdots \\ cv_k \end{bmatrix}\right)$$

$$= (cv_1)\mathbf{u}_1 + (cv_2)\mathbf{u}_2 + \cdots + (cv_k)\mathbf{u}_k$$
$$= c(v_1\mathbf{u}_1 + v_2\mathbf{u}_2 + \cdots + v_k\mathbf{u}_k)$$
$$= cT(\mathbf{v}).$$

Thus T is a linear transformation.

(b)   Since $\{\mathbf{u}_1, \mathbf{u}_2, \ldots, \mathbf{u}_k\}$ is linearly independent,

$$x_1\mathbf{u}_1 + x_2\mathbf{u}_2 + \cdots + x_k\mathbf{u}_k = \mathbf{0}$$

implies $x_1 = x_2 = \cdots = x_k = 0$. Thus $T(\mathbf{x}) = \mathbf{0}$ implies $\mathbf{x} = \mathbf{0}$, so that the null space of T is $\{\mathbf{0}\}$. It follows from Theorem 2.13 that T is one-to-one.

(c)   For every $\mathbf{x}$ in $\mathcal{R}^k$, $T(\mathbf{x})$ is a linear combination of $\mathbf{u}_1, \mathbf{u}_2, \ldots, \mathbf{u}_k$ and hence is a vector in V. Conversely, if $\mathbf{v}$ is in V, then $\mathbf{v} = a_1\mathbf{u}_1 + a_2\mathbf{u}_2 + \cdots + a_k\mathbf{u}_k$ for some scalars $a_1, a_2, \ldots, a_k$. For

$$\mathbf{a} = \begin{bmatrix} a_1 \\ a_2 \\ \vdots \\ a_k \end{bmatrix},$$

in $\mathcal{R}^k$, we have $T(\mathbf{a}) = \mathbf{v}$. Hence every vector in V is the image of a vector in $\mathcal{R}^k$. Thus the range of T is V.

49.   (a)   Let B be a 4×4 matrix such that $AB = O$. Then

$$O = AB = A[\mathbf{b}_1 \ \mathbf{b}_2 \ \mathbf{b}_3 \ \mathbf{b}_4] = [A\mathbf{b}_1 \ A\mathbf{b}_2 \ A\mathbf{b}_3 \ A\mathbf{b}_4].$$

So each column of B is a solution to $A\mathbf{x} = \mathbf{0}$. The reduced row echelon form of A is

$$\begin{bmatrix} 1 & 0 & -1 & -2 \\ 0 & 1 & 1 & -1 \\ 0 & 0 & 0 & 0 \\ 0 & 0 & 0 & 0 \end{bmatrix},$$

and so the parametric representation of the general solution of $A\mathbf{x} = \mathbf{0}$ is

$$\begin{bmatrix} x_1 \\ x_2 \\ x_3 \\ x_4 \end{bmatrix} = x_3 \begin{bmatrix} 1 \\ -1 \\ 1 \\ 0 \end{bmatrix} + x_4 \begin{bmatrix} 2 \\ 1 \\ 0 \\ 1 \end{bmatrix}.$$

Hence

$$B = \begin{bmatrix} 1 & 2 & 0 & 0 \\ -1 & 1 & 0 & 0 \\ 1 & 0 & 0 & 0 \\ 0 & 1 & 0 & 0 \end{bmatrix}$$

is a 4×4 matrix with rank 2 such that $AB = O$.

(b) If C is a 4×4 matrix such that $AC = O$, then the preceding argument shows that each column of C is a vector in Null A, a 2-dimensional subspace. Hence C can have at most two linearly independent columns, so that rank $C \leq 2$.

4.4   COORDINATE SYSTEMS

1. (a) False, every vector in V can be *uniquely* represented as a linear combination of the vectors in S only if S is a basis for V.
   (b) True
   (c) True
   (d) True
   (e) True

5. $\mathbf{v} = 2 \begin{bmatrix} 1 \\ -1 \end{bmatrix} + 5 \begin{bmatrix} -1 \\ 2 \end{bmatrix} = \begin{bmatrix} -3 \\ 8 \end{bmatrix}$

9. $\mathbf{v} = (-1) \begin{bmatrix} 0 \\ 1 \\ 1 \end{bmatrix} + 5 \begin{bmatrix} -1 \\ 0 \\ 1 \end{bmatrix} + (-2) \begin{bmatrix} 1 \\ 1 \\ 1 \end{bmatrix} = \begin{bmatrix} -7 \\ -3 \\ 2 \end{bmatrix}$

13. (a) Let B be the matrix whose columns are the vectors in $\mathcal{B}$. Since the reduced row echelon form of B is I_3, $\mathcal{B}$ is linearly independent. So $\mathcal{B}$ is a linearly independent set of 3 vectors from $\mathcal{R}^3$, and hence $\mathcal{B}$ is a basis for $\mathcal{R}^3$ by Theorem 4.5.
    (b) The components of $[\mathbf{v}]_\mathcal{B}$ are the coefficients that express $\mathbf{v}$ as a linear combination of the vectors in $\mathcal{B}$. Thus

$$[\mathbf{v}]_\mathcal{B} = \begin{bmatrix} 3 \\ 0 \\ -1 \end{bmatrix}.$$

17. By Theorem 4.9, $[\mathbf{v}]_\mathcal{B} = \begin{bmatrix} 1 & -1 \\ -1 & 2 \end{bmatrix}^{-1} \begin{bmatrix} 5 \\ -3 \end{bmatrix} = \begin{bmatrix} 7 \\ 2 \end{bmatrix}$.

21. By Theorem 4.9, $[\mathbf{v}]_\mathcal{B} = \begin{bmatrix} 0 & -1 & 1 \\ 1 & 0 & 1 \\ 1 & 1 & 1 \end{bmatrix}^{-1} \begin{bmatrix} 1 \\ -3 \\ -2 \end{bmatrix} = \begin{bmatrix} -5 \\ 1 \\ 2 \end{bmatrix}$.

25. (a) Since the reduced row echelon form of $\begin{bmatrix} 1 & 2 \\ 2 & 3 \end{bmatrix}$ is $\begin{bmatrix} 1 & 0 \\ 0 & 1 \end{bmatrix}$, $\mathcal{B}$ is a linearly independent subset of $\mathcal{R}^2$ containing 2 vectors. Hence $\mathcal{B}$ is a basis for $\mathcal{R}^2$ by Theorem 4.5.

(b) Let $B = [\mathbf{b}_1\ \mathbf{b}_2]$. Then

$$[\mathbf{e}_1]_\mathcal{B} = B^{-1}\mathbf{e}_1 = \begin{bmatrix} -3 \\ 2 \end{bmatrix} \qquad \text{and} \qquad [\mathbf{e}_2]_\mathcal{B} = B^{-1}\mathbf{e}_2 = \begin{bmatrix} 2 \\ -1 \end{bmatrix}.$$

Hence

$$A = \begin{bmatrix} -3 & 2 \\ 2 & -1 \end{bmatrix}.$$

(c) The matrices A and B are inverses of each other.

29. Let $\mathbf{v} = \begin{bmatrix} x \\ y \end{bmatrix}$ and $[\mathbf{v}]_\mathcal{B} = \begin{bmatrix} x' \\ y' \end{bmatrix}$, where $\mathcal{B}$ is the basis obtained by rotating the vectors in the standard basis by $30°$. As on page 234, we have

$$\begin{bmatrix} x' \\ y' \end{bmatrix} = [\mathbf{v}]_\mathcal{B} = (A_{30°})^{-1}\mathbf{v} = A_{30°}^T \mathbf{v} = \begin{bmatrix} \frac{\sqrt{3}}{2} & \frac{1}{2} \\ -\frac{1}{2} & \frac{\sqrt{3}}{2} \end{bmatrix} \begin{bmatrix} x \\ y \end{bmatrix}.$$

Hence

$$x' = \tfrac{\sqrt{3}}{2}x + \tfrac{1}{2}y$$
$$y' = -\tfrac{1}{2}x + \tfrac{\sqrt{3}}{2}y.$$

33. Let

$$B = \begin{bmatrix} 1 & 1 & 0 \\ 0 & 1 & -2 \\ 1 & 0 & 1 \end{bmatrix}.$$

Then, as in Exercise 29,

$$\begin{bmatrix} x' \\ y' \\ z' \end{bmatrix} = [\mathbf{v}]_\mathcal{B} = B^{-1}\mathbf{v} = \begin{bmatrix} -1 & 1 & 2 \\ 2 & -1 & -2 \\ 1 & -1 & -1 \end{bmatrix} \begin{bmatrix} x \\ y \\ z \end{bmatrix}.$$

Hence

$$x' = -x + y + 2z$$
$$y' = 2x - y - 2z$$
$$z' = x - y - z.$$

37. Let $\mathbf{v} = \begin{bmatrix} x \\ y \end{bmatrix}$ and $[\mathbf{v}]_{\mathcal{B}} = \begin{bmatrix} x' \\ y' \end{bmatrix}$, where $\mathcal{B}$ is the basis obtained by rotating the vectors in the standard basis by $60°$. As in Example 4,

$$\begin{bmatrix} x \\ y \end{bmatrix} = \mathbf{v} = A_{60°}[\mathbf{v}]_{\mathcal{B}} = \begin{bmatrix} \frac{1}{2} & -\frac{\sqrt{3}}{2} \\ \frac{\sqrt{3}}{2} & \frac{1}{2} \end{bmatrix} \begin{bmatrix} x' \\ y' \end{bmatrix}.$$

Hence

$$x = \tfrac{1}{2}x' - \tfrac{\sqrt{3}}{2}y'$$
$$y = \tfrac{\sqrt{3}}{2}x' + \tfrac{1}{2}y'.$$

41. Let

$$\begin{bmatrix} 1 & -1 & 0 \\ 3 & 1 & -1 \\ 0 & 1 & 1 \end{bmatrix}.$$

As in Exercise 37, we have

$$\begin{bmatrix} x \\ y \\ z \end{bmatrix} = \mathbf{v} = B[\mathbf{v}]_{\mathcal{B}} = \begin{bmatrix} 1 & -1 & 0 \\ 3 & 1 & -1 \\ 0 & 1 & 1 \end{bmatrix} \begin{bmatrix} x' \\ y' \\ z' \end{bmatrix}.$$

Thus

$$x = x' - y'$$
$$y = 3x' + y' - z'$$
$$z = y' + z'.$$

45. As in Exercise 29, we have

$$\begin{bmatrix} x' \\ y' \end{bmatrix} = A_{60°}^{T} \begin{bmatrix} x \\ y \end{bmatrix} = \begin{bmatrix} \frac{1}{2} & \frac{\sqrt{3}}{2} \\ -\frac{\sqrt{3}}{2} & \frac{1}{2} \end{bmatrix} \begin{bmatrix} x \\ y \end{bmatrix}.$$

Thus

$$x' = \tfrac{1}{2}x + \tfrac{\sqrt{3}}{2}y$$
$$y' = -\tfrac{\sqrt{3}}{2}x + \tfrac{1}{2}y.$$

Rewrite the given equation in the form

$$25(x')^2 + 16(y')^2 = 400.$$

Then substitute the expressions for x' and y' into this equation to obtain

$$\frac{73}{4}x^2 + \frac{9\sqrt{3}}{2}xy + \frac{91}{4}y^2 = 400,$$

that is,

$$73x^2 + 18\sqrt{3}xy + 91y^2 = 1600.$$

49. As in Exercise 37, we have

$$\begin{bmatrix} x \\ y \end{bmatrix} = A_{45^\circ} \begin{bmatrix} x' \\ y' \end{bmatrix} = \begin{bmatrix} \frac{\sqrt{2}}{2} & -\frac{\sqrt{2}}{2} \\ \frac{\sqrt{2}}{2} & \frac{\sqrt{2}}{2} \end{bmatrix} \begin{bmatrix} x' \\ y' \end{bmatrix}.$$

Thus

$$x = \frac{\sqrt{2}}{2}x' - \frac{\sqrt{2}}{2}y'$$
$$y = \frac{\sqrt{2}}{2}x' + \frac{\sqrt{2}}{2}y'.$$

Substituting these expressions for x and y into the given equation produces

$$4(x')^2 - 10(y')^2 = 20,$$

that is,

$$2(x')^2 - 5(y')^2 = 10.$$

53. By the definition of $[\mathbf{v}]_A$, we have

$$\mathbf{v} = a_1\mathbf{u}_1 + a_2\mathbf{u}_2 + \cdots + a_n\mathbf{u}_n$$
$$= \frac{a_1}{c_1}(c_1\mathbf{u}_1) + \frac{a_2}{c_2}(c_2\mathbf{u}_2) + \cdots + \frac{a_n}{c_n}(c_n\mathbf{u}_n).$$

Hence

$$[\mathbf{v}]_{\mathcal{B}} = \begin{bmatrix} \frac{a_1}{c_1} \\ \frac{a_2}{c_2} \\ \vdots \\ \frac{a_n}{c_n} \end{bmatrix}.$$

57. Consider $\mathcal{A} = \{\mathbf{e}_1, \mathbf{e}_2\}$ and $\mathcal{B} = \{\mathbf{e}_1, 2\mathbf{e}_2\}$. Then

$$[\mathbf{e}_1]_{\mathcal{A}} = \mathbf{e}_1 \qquad \text{and} \qquad [\mathbf{e}_1]_{\mathcal{B}} = \mathbf{e}_1,$$

but $\mathcal{A} \neq \mathcal{B}$.

61. **(a)** Let B be the matrix whose columns are the vectors in $\mathcal{B}$. By Theorem 4.9,

$$T(\mathbf{v}) = [\mathbf{v}]_{\mathcal{B}} = B^{-1}\mathbf{v}$$

for every vector $\mathbf{v}$ in $\mathcal{R}^n$. Hence T is the matrix transformation induced by B^{-1}, and so T is a linear transformation.

(b) If $T(\mathbf{v}) = \mathbf{0}$, then $B^{-1}\mathbf{v} = \mathbf{0}$. Hence $\mathbf{v} = \mathbf{0}$; so T is one-to-one. Let $\mathbf{w}$ be any vector in $\mathcal{R}^n$. Then

$$T(B\mathbf{w}) = B^{-1}(B\mathbf{w}) = B^{-1}B\mathbf{w} = I_n\mathbf{w} = \mathbf{w}.$$

So $\mathbf{w}$ is the image of a vector $B\mathbf{w}$ in $\mathcal{R}^n$. Therefore T is onto.

65. Suppose that $\mathcal{A} = \{\mathbf{u}_1, \mathbf{u}_2, \ldots, \mathbf{u}_k\}$ is a linearly independent subset of $\mathcal{R}^n$, and let $c_1, c_2, \ldots, c_k$ be scalars such that

$$c_1[\mathbf{u}_1]_{\mathcal{B}} + c_2[\mathbf{u}_2]_{\mathcal{B}} + \cdots + c_k[\mathbf{u}_k]_{\mathcal{B}} = \mathbf{0}.$$

Define $T: \mathcal{R}^n \to \mathcal{R}^n$ by $T(\mathbf{v}) = [\mathbf{v}]_{\mathcal{B}}$ for all $\mathbf{v}$ in $\mathcal{R}^n$. Then T is a linear transformation by Exercise 61(a), and so

$$c_1 T(\mathbf{u}_1) + c_2 T(\mathbf{u}_2) + \cdots + c_k T(\mathbf{u}_k) = \mathbf{0}$$
$$T(c_1\mathbf{u}_1 + c_2\mathbf{u}_2 + \cdots + c_k\mathbf{u}_k) = \mathbf{0}.$$

Thus $c_1\mathbf{u}_1 + c_2\mathbf{u}_2 + \cdots + c_k\mathbf{u}_k$ is in the null space of T. Since T is one-to-one by Exercise 61(b), it follows that $c_1\mathbf{u}_1 + c_2\mathbf{u}_2 + \cdots + c_k\mathbf{u}_k = \mathbf{0}$. Hence the linear independence of $\{\mathbf{u}_1, \mathbf{u}_2, \ldots, \mathbf{u}_k\}$ yields $c_1 = c_2 = \cdots = c_k = 0$. Therefore $\{[\mathbf{u}_1]_{\mathcal{B}}, [\mathbf{u}_2]_{\mathcal{B}}, \ldots, [\mathbf{u}_k]_{\mathcal{B}}\}$ is linearly independent.

The proof of the converse is similar.

69. Let B be the matrix whose columns are the vectors in $\mathcal{B}$. Since $[\mathbf{v}]_{\mathcal{B}} = B^{-1}\mathbf{v}$, we must find a nonzero vector $\mathbf{v}$ in $\mathcal{R}^5$ such that

$$B^{-1}\mathbf{v} = .5\mathbf{v}$$
$$B^{-1}\mathbf{v} - .5\mathbf{v} = \mathbf{0}$$
$$(B^{-1} - .5I_5)\mathbf{v} = \mathbf{0}.$$

The reduced row echelon form of $B^{-1} - .5I_5$ is

$$\begin{bmatrix} 1 & 0 & 0 & 0 & 0 \\ 0 & 1 & 0 & 0 & -2 \\ 0 & 0 & 1 & 0 & 2 \\ 0 & 0 & 0 & 1 & -2 \\ 0 & 0 & 0 & 0 & 0 \end{bmatrix}.$$

Thus the parametric representation of the general solution to $(B^{-1} - .5I_5)\mathbf{x} = \mathbf{0}$ is

$$\begin{bmatrix} x_1 \\ x_2 \\ x_3 \\ x_4 \\ x_5 \end{bmatrix} = x_5 \begin{bmatrix} 0 \\ 2 \\ -2 \\ 2 \\ 1 \end{bmatrix}.$$

So by taking

$$\mathbf{v} = \begin{bmatrix} 0 \\ 2 \\ -2 \\ 2 \\ 1 \end{bmatrix},$$

we have $[\mathbf{v}]_\mathcal{B} = .5\mathbf{v}$.

4.5   MATRIX REPRESENTATIONS OF LINEAR OPERATORS

1. **(a)** True

   (b) False, the matrix representation of T with respect to $\mathcal{B}$ is

   $$[[T(\mathbf{b}_1)]_\mathcal{B} \ [T(\mathbf{b}_2)]_\mathcal{B} \ \cdots \ [T(\mathbf{b}_n)]_\mathcal{B}]$$

   (c) True

   (d) False, $[T]_\mathcal{B} = B^{-1}AB$.

   (e) True

   (f) True

5. Since $T(\mathbf{b}_1) = 0\mathbf{b}_1 - 5\mathbf{b}_2 + 4\mathbf{b}_3$, we have

   $$[T(\mathbf{b}_1)]_\mathcal{B} = \begin{bmatrix} 0 \\ -5 \\ 4 \end{bmatrix}.$$

   Likewise

   $$[T(\mathbf{b}_2)]_\mathcal{B} = \begin{bmatrix} 2 \\ 0 \\ -7 \end{bmatrix} \quad \text{and} \quad [T(\mathbf{b}_2)]_\mathcal{B} = \begin{bmatrix} 3 \\ 0 \\ 1 \end{bmatrix}.$$

   Hence

   $$[T]_\mathcal{B} = \begin{bmatrix} 0 & 2 & 3 \\ -5 & 0 & 0 \\ 4 & -7 & 1 \end{bmatrix}.$$

9. Let

$$\mathbf{b}_1 = \begin{bmatrix} 1 \\ 0 \\ 1 \end{bmatrix}, \qquad \mathbf{b}_2 = \begin{bmatrix} 0 \\ 1 \\ 0 \end{bmatrix}, \qquad \text{and} \qquad \mathbf{b}_3 = \begin{bmatrix} 1 \\ 1 \\ 0 \end{bmatrix}.$$

(a) Since $T(\mathbf{b}_1) = 0\mathbf{b}_1 - \mathbf{b}_2 + 0\mathbf{b}_3$, we have

$$[T(\mathbf{b}_1)]_\mathcal{B} = \begin{bmatrix} 0 \\ -1 \\ 0 \end{bmatrix}.$$

Likewise

$$[T(\mathbf{b}_2)]_\mathcal{B} = \begin{bmatrix} 0 \\ 0 \\ 2 \end{bmatrix} \qquad \text{and} \qquad [T(\mathbf{b}_3)]_\mathcal{B} = \begin{bmatrix} 1 \\ 2 \\ 0 \end{bmatrix}.$$

Hence

$$[T]_\mathcal{B} = \begin{bmatrix} 0 & 0 & 1 \\ -1 & 0 & 2 \\ 0 & 2 & 0 \end{bmatrix}.$$

(b) The standard matrix A of T is given by

$$A = B[T]_\mathcal{B}B^{-1} = \begin{bmatrix} -1 & 2 & 1 \\ 0 & 2 & -1 \\ 1 & 0 & -1 \end{bmatrix},$$

where $B = [\mathbf{b}_1 \ \mathbf{b}_2 \ \mathbf{b}_3]$.

(c) For any vector $\mathbf{x}$ in $\mathcal{R}^3$, we have

$$T(\mathbf{x}) = A\mathbf{x} = \begin{bmatrix} -1 & 2 & 1 \\ 0 & 2 & -1 \\ 1 & 0 & -1 \end{bmatrix} \begin{bmatrix} x_1 \\ x_2 \\ x_3 \end{bmatrix} = \begin{bmatrix} -x_1 + 2x_2 + x_3 \\ 2x_2 - x_3 \\ x_1 - x_3 \end{bmatrix}.$$

13. The standard matrix of T is

$$\begin{bmatrix} 1 & 2 \\ 1 & 1 \end{bmatrix}.$$

If B is the matrix whose columns are the vectors in $\mathcal{B}$, then, by Theorem 4.10, we have

$$[T]_\mathcal{B} = B^{-1}AB = \begin{bmatrix} 1 & 2 \\ 1 & 1 \end{bmatrix}.$$

17. The standard matrix of T is

$$A = \begin{bmatrix} 0 & 4 & 0 \\ 1 & 0 & 2 \\ 0 & -2 & 3 \end{bmatrix}.$$

If B is the matrix whose columns are the vectors in $\mathcal{B}$, then, by Theorem 4.10, we have

$$[T]_{\mathcal{B}} = B^{-1}AB = \begin{bmatrix} 0 & -19 & 28 \\ 3 & 34 & -47 \\ 3 & 23 & -31 \end{bmatrix}.$$

21. Let A be the standard matrix of T and B be the matrix whose columns are the vectors in $\mathcal{B}$. Then, by Theorem 4.10, we have

$$A = B[T]_{\mathcal{B}}B^{-1} = \begin{bmatrix} 10 & -19 \\ 3 & -4 \end{bmatrix}.$$

25. From Exercise 3, we have

$$[T]_{\mathcal{B}} = \begin{bmatrix} 1 & -3 \\ 4 & 0 \end{bmatrix}.$$

Hence, by Theorem 4.11,

$$[T(3\mathbf{b}_1 - 2\mathbf{b}_2)]_{\mathcal{B}} = [T]_{\mathcal{B}}[3\mathbf{b}_1 - 2\mathbf{b}_2]_{\mathcal{B}} = [T]_{\mathcal{B}}\begin{bmatrix} 3 \\ -2 \end{bmatrix} = \begin{bmatrix} 9 \\ 12 \end{bmatrix}.$$

Therefore $T(3\mathbf{b}_1 - 2\mathbf{b}_2) = 9\mathbf{b}_1 + 12\mathbf{b}_2$.

29. For any $\mathbf{v}$ in $\mathcal{R}^n$, we have $I(\mathbf{v}) = \mathbf{v}$. Hence if $\mathcal{B} = \{\mathbf{b}_1, \mathbf{b}_2, \ldots, \mathbf{b}_n\}$, then

$$\begin{aligned} I_{\mathcal{B}} &= [[I(\mathbf{b}_1)]_{\mathcal{B}} \ [I(\mathbf{b}_2)]_{\mathcal{B}} \ \cdots \ [I(\mathbf{b}_n)]_{\mathcal{B}}] \\ &= [[\mathbf{b}_1]_{\mathcal{B}} \ [\mathbf{b}_2]_{\mathcal{B}} \ \cdots \ [\mathbf{b}_n]_{\mathcal{B}}] \\ &= [\mathbf{e}_1 \ \mathbf{e}_2 \ \cdots \ \mathbf{e}_n] \\ &= I_n. \end{aligned}$$

33. Take

$$\mathbf{b}_1 = \begin{bmatrix} 1 \\ -2 \end{bmatrix} \quad \text{and} \quad \mathbf{b}_2 = \begin{bmatrix} 2 \\ 1 \end{bmatrix}.$$

Then $\mathbf{b}_1$ lies on the line with equation $y = -2x$, and $\mathbf{b}_2$ is perpendicular to this line. Hence if $\mathcal{B} = \{\mathbf{b}_1, \mathbf{b}_2\}$, then

$$[T]_{\mathcal{B}} = \begin{bmatrix} 1 & 0 \\ 0 & -1 \end{bmatrix}.$$

So the standard matrix of T is

$$B[T]_{\mathcal{B}}B^{-1} = \begin{bmatrix} -.6 & -.8 \\ -.8 & .6 \end{bmatrix},$$

where $B = [\mathbf{b}_1 \ \mathbf{b}_2]$. Thus

$$T\left(\begin{bmatrix} x_1 \\ x_2 \end{bmatrix}\right) = \begin{bmatrix} -.6x_1 - .8x_2 \\ -.8x_1 + .6x_2 \end{bmatrix}.$$

37. Take

$$\mathbf{b}_1 = \begin{bmatrix} 1 \\ -3 \end{bmatrix} \quad \text{and} \quad \mathbf{b}_2 = \begin{bmatrix} 3 \\ 1 \end{bmatrix}.$$

Then $\mathbf{b}_1$ lies on the line with equation $y = -3x$, and $\mathbf{b}_2$ is perpendicular to this line. Hence if $\mathcal{B} = \{\mathbf{b}_1, \mathbf{b}_2\}$, then

$$[U]_{\mathcal{B}} = \begin{bmatrix} 1 & 0 \\ 0 & 0 \end{bmatrix}.$$

It follows that the standard matrix of U is

$$B[U]_{\mathcal{B}}B^{-1} = \begin{bmatrix} .1 & -.3 \\ -.3 & .9 \end{bmatrix},$$

where $B = [\mathbf{b}_1 \ \mathbf{b}_2]$. Therefore

$$U\left(\begin{bmatrix} x_1 \\ x_2 \end{bmatrix}\right) = \begin{bmatrix} .1x_1 - .3x_2 \\ -.3x_1 + .9x_2 \end{bmatrix}.$$

41. Let

$$\mathbf{b}_1 = \begin{bmatrix} -2 \\ 1 \\ 0 \end{bmatrix}, \quad \mathbf{b}_2 = \begin{bmatrix} 3 \\ 0 \\ 1 \end{bmatrix}, \quad \text{and} \quad \mathbf{b}_3 = \begin{bmatrix} 1 \\ 2 \\ -3 \end{bmatrix}.$$

(a) Since $\mathbf{b}_1$ and $\mathbf{b}_2$ lie in W, we have $U(\mathbf{b}_1) = \mathbf{b}_1$ and $U(\mathbf{b}_2) = \mathbf{b}_2$. Moreover, since $\mathbf{b}_3$ is perpendicular to W, $U(\mathbf{b}_3) = \mathbf{0}$. Thus

$$[U]_{\mathcal{B}} = [[U(\mathbf{b}_1)]_{\mathcal{B}} \ [U(\mathbf{b}_2)]_{\mathcal{B}} \ [U(\mathbf{b}_3)]_{\mathcal{B}}] = [\mathbf{e}_1 \ \mathbf{e}_2 \ \mathbf{0}] = \begin{bmatrix} 1 & 0 & 0 \\ 0 & 1 & 0 \\ 0 & 0 & 0 \end{bmatrix}.$$

(b) Let $B = [\mathbf{b}_1 \ \mathbf{b}_2 \ \mathbf{b}_3]$. Then by Theorem 4.10, the standard matrix of U is

$$B[U]_{\mathcal{B}}B^{-1} = \frac{1}{14}\begin{bmatrix} 13 & -2 & 3 \\ -2 & 10 & 6 \\ 3 & 6 & 5 \end{bmatrix}.$$

(c)   Using the preceding standard matrix of U, we have

$$U\left(\begin{bmatrix} x_1 \\ x_2 \\ x_3 \end{bmatrix}\right) = \frac{1}{14}\begin{bmatrix} 13x_1 - 2x_2 + 3x_3 \\ -2x_1 + 10x_2 + 6x_3 \\ 3x_1 + 6x_2 + 5x_3 \end{bmatrix}.$$

45.   Let A be the standard matrix of T and B be the matrix whose columns are the vectors in $\mathcal{B}$. Then

$$[T]_{\mathcal{B}} = B^{-1}AB$$

by Theorem 4.10. Since B and B^{-1} are invertible, $\operatorname{rank}[T]_{\mathcal{B}} = \operatorname{rank} A$ by Theorem 2.7. Because the range of T equals the column space of A, the dimension of the range of T equals the dimension of the column space of A, which is $\operatorname{rank} A$. Thus the dimension of the range of T equals $\operatorname{rank}[T]_{\mathcal{B}}$.

49.   Let A and B be the matrices whose columns are the vectors in $\mathcal{A}$ and $\mathcal{B}$, respectively, and let C be the standard matrix of T. Then

$$[T]_{\mathcal{A}} = A^{-1}CA \qquad \text{and} \qquad [T]_{\mathcal{B}} = B^{-1}CB$$

by Theorem 4.10. Solving the second equation for C, we obtain $C = B[T]_{\mathcal{B}}B^{-1}$. Hence

$$\begin{aligned} [T]_{\mathcal{A}} = A^{-1}CA &= A^{-1}(B[T]_{\mathcal{B}}B^{-1})A \\ &= (A^{-1}B)[T]_{\mathcal{B}}(B^{-1}A) \\ &= (B^{-1}A)^{-1}[T]_{\mathcal{B}}(B^{-1}A) \end{aligned}$$

by Theorem 2.3(b). Thus $[T]_{\mathcal{A}}$ and $[T]_{\mathcal{B}}$ are similar.

53.   (a)   Let $B = [\mathbf{b}_1\ \mathbf{b}_2\ \mathbf{b}_3\ \mathbf{b}_4]$. The standard matrices of T and U are

$$A = \begin{bmatrix} 1 & -2 & 0 & 0 \\ 0 & 0 & 1 & 0 \\ -1 & 0 & 3 & 0 \\ 0 & 2 & 0 & -1 \end{bmatrix} \qquad \text{and} \qquad C = \begin{bmatrix} 0 & 1 & -1 & 2 \\ -2 & 0 & 0 & 3 \\ 0 & 2 & -1 & 0 \\ 3 & 0 & 0 & 1 \end{bmatrix},$$

respectively. Hence, by Theorem 2.14, the standard matrix of UT is

$$CA = \begin{bmatrix} 1 & 4 & -2 & -2 \\ -2 & 10 & 0 & -3 \\ 1 & 0 & -1 & 0 \\ 3 & -4 & 0 & -1 \end{bmatrix}.$$

Thus, by Theorem 4.10, we have

$$[T]_{\mathcal{B}} = B^{-1}AB = \begin{bmatrix} 11 & 5 & 13 & 1 \\ -2 & 0 & -5 & -3 \\ -8 & -3 & -9 & 0 \\ 6 & 1 & 8 & 1 \end{bmatrix},$$

$$[U]_{\mathcal{B}} = B^{-1}CB = \begin{bmatrix} -5 & 10 & -38 & -31 \\ 2 & -3 & 9 & 6 \\ 6 & -10 & 27 & 17 \\ -4 & 7 & -25 & -19 \end{bmatrix},$$

and

$$[UT]_{\mathcal{B}} = B^{-1}(CA)B = \begin{bmatrix} 43 & 58 & -21 & -66 \\ -8 & -11 & 8 & 17 \\ -28 & -34 & 21 & 53 \\ 28 & 36 & -14 & -44 \end{bmatrix}.$$

(b) From (a), we see that

$$[U]_{\mathcal{B}}[T]_{\mathcal{B}} = (B^{-1}CB)(B^{-1}AB)$$
$$= B^{-1}CI_4AB$$
$$= B^{-1}CAB$$
$$= [UT]_{\mathcal{B}}.$$

57. We claim that $[T^{-1}]_{\mathcal{B}} = ([T]_{\mathcal{B}})^{-1}$. By Theorem 4.10,

$$[T]_{\mathcal{B}} = B^{-1}AB,$$

where A is the standard matrix of T and B is the matrix whose columns are the vectors in $\mathcal{B}$. Recall from Theorem 2.14(b) that A is invertible and that the standard matrix of T^{-1} is A^{-1}. Hence, by Theorems 4.10 and 2.3(b), we have

$$[T^{-1}]_{\mathcal{B}} = B^{-1}A^{-1}B = (B^{-1}AB)^{-1} = [T]_{\mathcal{B}}^{-1}.$$

CHAPTER 4 REVIEW

1. (a) True
  (b) True
  (c) False, the null space of an $m \times n$ matrix is contained in $\mathcal{R}^n$.
  (d) False, the column space of an $m \times n$ matrix is contained in $\mathcal{R}^m$.
  (e) False, the row space of an $m \times n$ matrix is contained in $\mathcal{R}^n$.
  (f) True
  (g) True
  (h) False, the range of every linear transformation equals the column space of its standard matrix.
  (i) False, a nonzero subspace of $\mathcal{R}^n$ has infinitely many bases.
  (j) False, every basis for a particular subspace contains the same number of vectors.

(k) True
(l) True
(m) True
(n) True
(o) True
(p) True
(q) False, the dimension of the null space of a matrix equals the *nullity* of the matrix.
(r) True
(s) False, the dimension of the row space of a matrix equals the *rank* of the matrix.
(t) False, consider $A = \begin{bmatrix} 1 & 2 \\ 1 & 2 \end{bmatrix}$.
(u) True
(v) True
(w) False, $[T]_\mathcal{B} = B^{-1}AB$.
(x) True

3. (a) By Theorem 4.3, a spanning set for V must contain a basis for V. Since every basis for V contains $\dim V = k$ vectors, every spanning set for V must contain at least k vectors.

   (b) By Theorem 1.9′ on page 213, every linearly independent subset of V contains at most k vectors.

5. Consider

$$\mathbf{u} = \begin{bmatrix} -1 \\ 0 \\ 1 \\ 0 \end{bmatrix} \qquad \text{and} \qquad \mathbf{v} = \begin{bmatrix} 1 \\ 0 \\ 1 \\ 0 \end{bmatrix}.$$

   Both $\mathbf{u}$ and $\mathbf{v}$ belong to the given set, but their sum

$$\mathbf{u} + \mathbf{v} = \begin{bmatrix} 0 \\ 0 \\ 2 \\ 0 \end{bmatrix}$$

   does not.

7. The reduced row echelon form of the given matrix A is

$$\begin{bmatrix} 1 & 0 & 3 \\ 0 & 1 & -2 \\ 0 & 0 & 0 \\ 0 & 0 & 0 \end{bmatrix}.$$

(a) The null space of A consists of the solutions to $A\mathbf{x} = \mathbf{0}$. The general solution of this system is

$$\begin{aligned} x_1 &= -3x_3 \\ x_2 &= 2x_3 \\ x_2 &\ \text{free.} \end{aligned}$$

Hence the parametric form of the general solution of $A\mathbf{x} = \mathbf{0}$ is

$$\begin{bmatrix} x_1 \\ x_2 \\ x_3 \end{bmatrix} = x_1 \begin{bmatrix} -3 \\ 2 \\ 1 \end{bmatrix}.$$

Thus

$$\left\{ \begin{bmatrix} -3 \\ 2 \\ 1 \end{bmatrix} \right\}$$

is a basis for the null space of A.

(b) The pivot columns of the given matrix form a basis for its column space. From R above, we see that the pivot columns of the given matrix are columns 1 and 2. Hence

$$\left\{ \begin{bmatrix} 1 \\ -1 \\ 2 \\ 1 \end{bmatrix}, \begin{bmatrix} 2 \\ -1 \\ 1 \\ 4 \end{bmatrix} \right\}$$

is a basis for the column space of the given matrix.

(c) A basis for the row space of the given matrix consists of the nonzero rows in its reduced row echelon form. From R above, we see that this basis is

$$\{[1\ \ 0\ \ 3],\ [0\ \ 1\ \ -2]\}.$$

9. The standard matrix of T is

$$A = \begin{bmatrix} 0 & 1 & -2 \\ -1 & 3 & 1 \\ 1 & -4 & 1 \\ 2 & -1 & 3 \end{bmatrix},$$

and its reduced row echelon form is $[\mathbf{e}_1\ \mathbf{e}_2\ \mathbf{e}_3]$.

(a) The set

$$\left\{ \begin{bmatrix} 0 \\ -1 \\ 1 \\ 2 \end{bmatrix}, \begin{bmatrix} 1 \\ 3 \\ -4 \\ -1 \end{bmatrix}, \begin{bmatrix} -2 \\ 1 \\ 1 \\ 3 \end{bmatrix} \right\}$$

of pivot columns of A is a basis for the range of T.

(b)   The only solution to $A\mathbf{x} = \mathbf{0}$ is $\mathbf{x} = \mathbf{0}$; so the null space of T is the zero subspace.

13.   Let B be the matrix whose columns are the vectors in $\mathcal{B}$.

(a)   Since the reduced row echelon form of B is I_3, $\mathcal{B}$ is a linearly independent subset of $\mathcal{R}^3$. Since $\mathcal{B}$ contains exactly 3 vectors, $\mathcal{B}$ is a basis for $\mathcal{R}^3$ by Theorem 4.5.

(b)   We have

$$\mathbf{v} = 4 \begin{bmatrix} 0 \\ -1 \\ 1 \end{bmatrix} - 3 \begin{bmatrix} 1 \\ 0 \\ -1 \end{bmatrix} - 2 \begin{bmatrix} -1 \\ -1 \\ 1 \end{bmatrix} = \begin{bmatrix} -1 \\ -2 \\ 5 \end{bmatrix}.$$

(c)   By Theorem 4.9, we have

$$[\mathbf{w}]_{\mathcal{B}} = B^{-1}\mathbf{w} = \begin{bmatrix} 1 \\ -8 \\ -6 \end{bmatrix}.$$

15.   (a)   Let B be the matrix whose columns are the vectors in $\mathcal{B}$. Then

$$[T(\mathbf{b}_1)]_{\mathcal{B}} = B^{-1}(T(\mathbf{b}_1)) = B^{-1} \begin{bmatrix} 3 \\ 4 \end{bmatrix} = \begin{bmatrix} -17 \\ -10 \end{bmatrix}$$

and

$$[T(\mathbf{b}_2)]_{\mathcal{B}} = B^{-1}(T(\mathbf{b}_2)) = B^{-1} \begin{bmatrix} -1 \\ 1 \end{bmatrix} = \begin{bmatrix} 1 \\ 1 \end{bmatrix}.$$

Therefore

$$[T]_{\mathcal{B}} = [[T(\mathbf{b}_1)]_{\mathcal{B}} \ \ [T(\mathbf{b}_2)]_{\mathcal{B}}] = \begin{bmatrix} -17 & 1 \\ -10 & 1 \end{bmatrix}.$$

(b)   Let A denote the standard matrix of T. By Theorem 4.10, we have

$$A = B[T]_{\mathcal{B}}B^{-1} = \begin{bmatrix} -7 & -5 \\ -14 & -9 \end{bmatrix}.$$

(c)   Using the result of (b), we have

$$T\left(\begin{bmatrix} x_1 \\ x_2 \end{bmatrix} \right) = A \begin{bmatrix} x_1 \\ x_2 \end{bmatrix} = \begin{bmatrix} -7x_1 - 5x_2 \\ -14x_1 - 9x_2 \end{bmatrix}.$$

17.   Let B be the matrix whose columns are the vectors in $\mathcal{B}$. By Theorem 4.10, the standard matrix of T is

$$B[T]_{\mathcal{B}}B^{-1} = \begin{bmatrix} 1 & 6 & -5 \\ -4 & 4 & 5 \\ -1 & 3 & 1 \end{bmatrix}.$$

Hence

$$T\left(\begin{bmatrix} x_1 \\ x_2 \\ x_3 \end{bmatrix}\right) = \begin{bmatrix} x_1 + 6x_2 - 5x_3 \\ -4x_1 + 4x_2 + 5x_3 \\ -x_1 + 3x_2 + x_3 \end{bmatrix}.$$

19. Let $\mathbf{v} = \begin{bmatrix} x \\ y \end{bmatrix}$ and $\mathbf{v}_\mathcal{B} = \begin{bmatrix} x' \\ y' \end{bmatrix}$, where $\mathcal{B}$ is the basis obtained by rotating the vectors in the standard basis by 120°. As on page 234, we have

$$\begin{bmatrix} x' \\ y' \end{bmatrix} = [\mathbf{v}]_\mathcal{B} = (A_{120°})^{-1}\mathbf{v} = (A_{120°})^T\mathbf{v} = \begin{bmatrix} -\frac{1}{2} & \frac{\sqrt{3}}{2} \\ -\frac{\sqrt{3}}{2} & -\frac{1}{2} \end{bmatrix}\begin{bmatrix} x \\ y \end{bmatrix}.$$

Hence

$$x' = -\tfrac{1}{2}x + \tfrac{\sqrt{3}}{2}y$$
$$y' = -\tfrac{\sqrt{3}}{2}x - \tfrac{1}{2}y.$$

Rewrite the given equation in the form

$$9(x')^2 + 4(y')^2 = 36,$$

and substitute the expressions above for x' and y'. The resulting equation is

$$\frac{21}{4}x^2 - \frac{5\sqrt{3}}{2}xy + \frac{31}{4}y^2 = 36,$$

that is,

$$21x^2 - 10\sqrt{3}xy + 31y^2 = 144.$$

21. Let

$$\mathbf{v} = \begin{bmatrix} x \\ y \end{bmatrix} \qquad \text{and} \qquad [\mathbf{v}]_\mathcal{B} = \begin{bmatrix} x' \\ y' \end{bmatrix},$$

where $\mathcal{B}$ is the basis obtained by rotating $\mathbf{e}_1$ and $\mathbf{e}_2$ through 315°. Then

$$\begin{bmatrix} x \\ y \end{bmatrix} = \mathbf{v} = A_{315°}[\mathbf{v}]_\mathcal{B} = \begin{bmatrix} \frac{\sqrt{2}}{2} & \frac{\sqrt{2}}{2} \\ -\frac{\sqrt{2}}{2} & \frac{\sqrt{2}}{2} \end{bmatrix}\begin{bmatrix} x' \\ y' \end{bmatrix}.$$

Hence

$$x = \tfrac{\sqrt{2}}{2}x' + \tfrac{\sqrt{2}}{2}y'$$
$$y = -\tfrac{\sqrt{2}}{2}x' + \tfrac{\sqrt{2}}{2}y'.$$

Substituting these expressions for x and y transforms $29x^2 - 42xy + 29y^2 = 200$ into the form

$$50(x')^2 + 8(y')^2 = 200.$$

23. Take

$$\mathbf{b}_1 = \begin{bmatrix} 2 \\ -3 \end{bmatrix} \quad \text{and} \quad \mathbf{b}_2 = \begin{bmatrix} 3 \\ 2 \end{bmatrix}.$$

Then $\mathbf{b}_1$ lies on the line with equation $y = -\frac{3}{2}x$, and $\mathbf{b}_2$ is perpendicular to this line. Hence if $\mathcal{B} = \{\mathbf{b}_1, \mathbf{b}_2\}$, then

$$[T]_\mathcal{B} = \begin{bmatrix} 1 & 0 \\ 0 & -1 \end{bmatrix}.$$

So the standard matrix of T is

$$A = B[T]_\mathcal{B}B^{-1} = \frac{1}{13} \begin{bmatrix} -5 & -12 \\ -12 & 5 \end{bmatrix},$$

where $B = [\mathbf{b}_1 \ \mathbf{b}_2]$. Thus

$$T\left(\begin{bmatrix} x_1 \\ x_2 \end{bmatrix}\right) = A\begin{bmatrix} x_1 \\ x_2 \end{bmatrix} = \frac{1}{13}\begin{bmatrix} -5x_1 - 12x_2 \\ -12x_1 + 5x_2 \end{bmatrix}.$$

25. (a) If S is linearly independent, then the number of vectors in S cannot exceed dim V. Thus $m \le k$.

(b) If S is a linearly dependent set, then there is no relationship between m and k, in general.

(c) If S is a spanning set for V, then the number of vectors in S must be at least the dimension of V. Thus $m \ge k$.

29. A vector in V is a solution to the system of linear equations

$$\begin{aligned} v_1 + v_2 \qquad &= 0 \\ 2v_1 \qquad - v_3 &= 0. \end{aligned}$$

The parametric representation of the general solution to this system is

$$\begin{bmatrix} v_1 \\ v_2 \\ v_3 \end{bmatrix} = v_3 \begin{bmatrix} 1 \\ -1 \\ 2 \end{bmatrix}.$$

Moreover, a vector in W is a solution to

$$\begin{aligned} w_1 \qquad - 2w_3 &= 0 \\ w_2 + 3w_3 &= 0. \end{aligned}$$

The parametric representation of the general solution to this system is

$$\begin{bmatrix} w_1 \\ w_2 \\ w_3 \end{bmatrix} = w_3 \begin{bmatrix} 2 \\ -1 \\ 1 \end{bmatrix}.$$

It follows that a vector in $V + W$ has the form

$$r \begin{bmatrix} 1 \\ -1 \\ 2 \end{bmatrix} + s \begin{bmatrix} 2 \\ -1 \\ 1 \end{bmatrix}$$

for some scalars r and s. So

$$\mathcal{B} = \left\{ \begin{bmatrix} 1 \\ -1 \\ 2 \end{bmatrix}, \begin{bmatrix} 2 \\ -1 \\ 1 \end{bmatrix} \right\}$$

is a spanning set for $V + W$. Moreover, $\mathcal{B}$ is linearly independent since neither vector in $\mathcal{B}$ is a multiple of the other. Since $\mathcal{B}$ is contained in $V + W$, $\mathcal{B}$ is a basis for $V + W$.

Chapter 5

Eigenvalues, Eigenvectors, and Diagonalization

5.1 EIGENVALUES AND EIGENVECTORS

1. (a) False, if $A\mathbf{v} = \lambda\mathbf{v}$ for some *nonzero* vector $\mathbf{v}$, then λ is an eigenvalue of A.
   (b) False, if $A\mathbf{v} = \lambda\mathbf{v}$ for some *nonzero* vector $\mathbf{v}$, then $\mathbf{v}$ is an eigenvector of A.
   (c) True
   (d) True
   (e) True
   (f) False, the eigenspace of A corresponding to eigenvalue λ is the null space of $A - \lambda I_n$.
   (g) True
   (h) True
   (i) False, the linear operator on $\mathcal{R}^2$ that rotates a vector by $90°$ has no real eigenvalues. (See page 256.)

5. The eigenvalue is -2 because $\begin{bmatrix} 19 & -7 \\ 42 & -16 \end{bmatrix} \begin{bmatrix} 1 \\ 3 \end{bmatrix} = \begin{bmatrix} -2 \\ -6 \end{bmatrix} = (-2) \begin{bmatrix} 1 \\ 3 \end{bmatrix}$.

9. The eigenvalue is -4 because

$$\begin{bmatrix} 2 & -6 & 6 \\ 1 & 9 & -6 \\ -2 & 16 & -13 \end{bmatrix} \begin{bmatrix} -1 \\ 1 \\ 2 \end{bmatrix} = \begin{bmatrix} 4 \\ -4 \\ -8 \end{bmatrix} = (-4) \begin{bmatrix} -1 \\ 1 \\ 2 \end{bmatrix}.$$

13. Let A denote the given matrix. The reduced row echelon form of $A - 3I_2$ is

$$\begin{bmatrix} 1 & 1 \\ 0 & 0 \end{bmatrix},$$

and so

$$\left\{ \begin{bmatrix} -1 \\ 1 \end{bmatrix} \right\}$$

is a basis for the null space of $A - 3I_2$, which is the eigenspace of A corresponding to eigenvalue 3.

17. Let A denote the given matrix. The reduced row echelon form of $A - 3I_3$ is

$$\begin{bmatrix} 1 & 1 & 0 \\ 0 & 0 & 1 \\ 0 & 0 & 0 \end{bmatrix},$$

and so

$$\left\{ \begin{bmatrix} -1 \\ 1 \\ 0 \end{bmatrix} \right\}$$

is a basis for the eigenspace of A corresponding to eigenvalue 3.

21. Let A denote the given matrix. The reduced row echelon form of $A - (-1)I_3 = A + I_3$ is

$$\begin{bmatrix} 1 & \frac{1}{3} & -\frac{2}{3} \\ 0 & 0 & 0 \\ 0 & 0 & 0 \end{bmatrix},$$

and so

$$\left\{ \begin{bmatrix} -1 \\ 3 \\ 0 \end{bmatrix}, \begin{bmatrix} 2 \\ 0 \\ 3 \end{bmatrix} \right\}$$

is a basis for the eigenspace corresponding to eigenvalue -1.

25. The eigenvalue is 6 because $T\left(\begin{bmatrix} -2 \\ 3 \end{bmatrix} \right) = \begin{bmatrix} -12 \\ 18 \end{bmatrix} = 6 \begin{bmatrix} -2 \\ 3 \end{bmatrix}.$

29. The eigenvalue is 5 because $T\left(\begin{bmatrix} -1 \\ 3 \\ 1 \end{bmatrix} \right) = \begin{bmatrix} -5 \\ 15 \\ 5 \end{bmatrix} = 5 \begin{bmatrix} -1 \\ 3 \\ 1 \end{bmatrix}.$

33. The standard matrix of T is

$$A = \begin{bmatrix} 1 & -1 & -3 \\ -3 & -1 & -9 \\ 1 & 1 & 5 \end{bmatrix},$$

and the reduced row echelon form of $A - 2I_3$ is

$$\begin{bmatrix} 1 & 1 & 3 \\ 0 & 0 & 0 \\ 0 & 0 & 0 \end{bmatrix}.$$

Hence the parametric representation of the general solution of $(A - 2I_3)\mathbf{x} = \mathbf{0}$ is

$$\begin{bmatrix} x_1 \\ x_2 \\ x_3 \end{bmatrix} = x_2 \begin{bmatrix} -1 \\ 1 \\ 0 \end{bmatrix} + x_3 \begin{bmatrix} -3 \\ 0 \\ 1 \end{bmatrix}.$$

Thus

$$\left\{ \begin{bmatrix} -1 \\ 1 \\ 0 \end{bmatrix}, \begin{bmatrix} -3 \\ 0 \\ 1 \end{bmatrix} \right\}$$

is a basis for the eigenspace corresponding to eigenvalue 2.

37. Since $I(\mathbf{v}) = \mathbf{v}$ for every $\mathbf{v}$ in $\mathcal{R}^n$, the only eigenvalue of I is 1. Moreover, the eigenspace of I corresponding to 1 is $\mathcal{R}^n$.

41. The eigenspace of A corresponding to 0 is the set of vectors $\mathbf{v}$ such that $A\mathbf{v} = 0\mathbf{v}$, that is, such that $A\mathbf{v} = \mathbf{0}$. So the eigenspace corresponding to 0 is the null space of A.

45. Let λ be an eigenvalue of A, and let $\mathbf{v}$ be an eigenvector of A corresponding to λ. Then $\mathbf{v} \neq \mathbf{0}$, and

$$A^2\mathbf{v} = A(A\mathbf{v}) = A(\lambda\mathbf{v}) = \lambda(A\mathbf{v}) = \lambda(\lambda\mathbf{v}) = \lambda^2\mathbf{v}.$$

Hence λ^2 is an eigenvalue of A^2.

49. Suppose that c_1 and c_2 are scalars such that

$$c_1\mathbf{v}_1 + c_2\mathbf{v}_2 = \mathbf{0}.$$

Then $c\mathbf{v}_1 = -c_2\mathbf{v}_2$, and

$$\begin{aligned} \mathbf{0} = T(\mathbf{0}) = T(c_1\mathbf{v}_1 + c_2\mathbf{v}_2) &= c_1T(\mathbf{v}_1) + c_2T(\mathbf{v}_2) \\ &= c_1\lambda_1\mathbf{v}_1 + c_2\lambda_2\mathbf{v}_2 \\ &= \lambda_1(-c_2\mathbf{v}_2) + c_2\lambda_2\mathbf{v}_2 \\ &= (\lambda_2 - \lambda_1)c_2\mathbf{v}_2. \end{aligned}$$

Now $\lambda_2 - \lambda_1 \neq 0$ and $\mathbf{v}_2 \neq \mathbf{0}$ (because $\mathbf{v}_2$ is an eigenvector of T). Hence $c_2 = 0$. It follows that $c_1\mathbf{v}_1 = \mathbf{0}$. Since $\mathbf{v}_1 \neq \mathbf{0}$, we also have $c_1 = 0$. Thus $\{\mathbf{v}_1, \mathbf{v}_2\}$ is linearly independent.

53. The eigenvalues of A are -2.7, 2.3, and -1.1 (with multiplicity 2), but the eigenvalues of $3A$ are -8.1, 6.9, and -3.3 (with multiplicity 2).

57. Yes, the eigenvalues of A^T are the same as those of A. Eigenvectors of A^T are found by solving $(A^T - \lambda I_4)\mathbf{x} = \mathbf{0}$ for each eigenvalue λ. Four eigenvectors of A^T are

$$\begin{bmatrix} -1 \\ 1 \\ -2 \\ 1 \end{bmatrix}, \quad \begin{bmatrix} 2 \\ 0 \\ 3 \\ 3 \end{bmatrix}, \quad \begin{bmatrix} 1 \\ -1 \\ 2 \\ 0 \end{bmatrix}, \quad \text{and} \quad \begin{bmatrix} 0 \\ -1 \\ 0 \\ 1 \end{bmatrix}.$$

5.2   THE CHARACTERISTIC POLYNOMIAL

1. **(a)** False, consider the matrix A in Example 1 and $B = \begin{bmatrix} -3 & 0 \\ 0 & 5 \end{bmatrix}$, which both have $(t+3)(t-5)$ as their characteristic polynomial.

(b) True

(c) True

(d) False, see page 260.

(e) False, see page 260.

(f) False, consider I_n.

(g) False, the rotation matrix $A_{90°}$ has no eigenvectors in $\mathcal{R}^2$.

(h) True

(i) False, $\begin{bmatrix} 0 & -1 \\ 1 & 0 \end{bmatrix}$ has a characteristic polynomial of $t^2 + 1$.

(j) True

(k) False, consider $4I_3$; here 4 is an eigenvalue of multiplicity 3.

(l) False, see Example 3.

3. **(a)** By Theorem 5.1, the eigenvalue 5 must have a multiplicity of 3 or more. In addition, the eigenvalue -9 must have a multiplicity of 1 or more. Since A is a 4×4 matrix, the sum of the multiplicities of its two eigenvalues must be 4. Hence eigenvalue 5 must have multiplicity 3, and eigenvalue -9 must have multiplicity 1. Thus the characteristic polynomial of A must be $(t-5)^3(t+9)$.

(b) By Theorem 5.1, the eigenvalue -9 must have a multiplicity of 1 or more. As in (a), the sum of the multiplicities of the two eigenvalues of A must be 4. Since eigenvalue 5 must have a multiplicity of at least one, there are three possibilities:

   (i) Eigenvalue 5 has multiplicity 1, and eigenvalue -9 has multiplicity 3, in which case the characteristic polynomial of A is $(t-5)(t+9)^3$.

  (ii) Eigenvalue 5 has multiplicity 2, and eigenvalue -9 has multiplicity 2, in which case the characteristic polynomial of A is $(t-5)^2(t+9)^2$.

 (iii) Eigenvalue 5 has multiplicity 3, and eigenvalue -9 has multiplicity 1, in which case the characteristic polynomial of A is $(t-5)^3(t+9)$.

(c) If dim $W_1 = 2$, then eigenvalue 5 must have a multiplicity of 2 or more. This leads to the two cases described in (ii) and (iii) of (b).

5. The eigenvalues of the matrix are the roots of the characteristic polynomial, which are 5 and 6.

The parametric representation of the general solution to $(A - 5I_2)\mathbf{x} = \mathbf{0}$ is

$$\begin{bmatrix} x_1 \\ x_2 \end{bmatrix} = x_2 \begin{bmatrix} -1.5 \\ 1 \end{bmatrix}.$$

So

$$\left\{ \begin{bmatrix} -1.5 \\ 1 \end{bmatrix} \right\} \quad \text{or} \quad \left\{ \begin{bmatrix} -3 \\ 2 \end{bmatrix} \right\}$$

is a basis for the eigenspace of A corresponding to eigenvalue 5.

The parametric representation of the general solution to $(A - 6I_2)\mathbf{x} = \mathbf{0}$ is

$$\begin{bmatrix} x_1 \\ x_2 \end{bmatrix} = x_2 \begin{bmatrix} -1 \\ 1 \end{bmatrix}.$$

Thus

$$\left\{ \begin{bmatrix} -1 \\ 1 \end{bmatrix} \right\}$$

is a basis for the eigenspace of A corresponding to eigenvalue 6.

9. The eigenvalues of the matrix are -3 and 2 (with multiplicity 2). As in Exercise 5, we find that a basis for each eigenspace is

$$\left\{ \begin{bmatrix} 1 \\ 1 \\ 1 \end{bmatrix} \right\} \quad \text{and} \quad \left\{ \begin{bmatrix} 1 \\ 0 \\ 1 \end{bmatrix} \right\}.$$

13. The eigenvalues of the matrix are -3, -2, and 1. As in Exercise 5, we find that a basis for each eigenspace is

$$\left\{ \begin{bmatrix} -1 \\ 1 \\ 1 \end{bmatrix} \right\}, \quad \left\{ \begin{bmatrix} -1 \\ 1 \\ 0 \end{bmatrix} \right\}, \quad \text{and} \quad \left\{ \begin{bmatrix} 1 \\ 0 \\ 1 \end{bmatrix} \right\}.$$

17. The characteristic polynomial of the matrix is $t^2 + 3t - 4 = (t + 4)(t - 1)$; so the eigenvalues are -4 and 1. Bases for the corresponding eigenspaces are found as in Exercise 5; they are

$$\left\{ \begin{bmatrix} 1 \\ 2 \end{bmatrix} \right\} \quad \text{and} \quad \left\{ \begin{bmatrix} 1 \\ 1 \end{bmatrix} \right\}.$$

21. The characteristic polynomial of the matrix is $-t^3 - 5t^2 - 3t + 9 = -(t-1)(t+3)^2$; so the eigenvalues are 1 and -3 (with multiplicity 2). Bases for the corresponding eigenspaces are found as in Exercise 5; they are

$$\left\{ \begin{bmatrix} 1 \\ 0 \\ 2 \end{bmatrix} \right\} \quad \text{and} \quad \left\{ \begin{bmatrix} 1 \\ 0 \\ 1 \end{bmatrix} \right\}.$$

25. The characteristic polynomial of the matrix is $-t^3 - 4t^2 + 20t + 48 = -(t+6)(t+2)(t-4)$; so the eigenvalues are -6, -2, and 4. Bases for the corresponding eigenspaces are found as in Exercise 5; they are

$$\left\{ \begin{bmatrix} -1 \\ 1 \\ 1 \end{bmatrix} \right\}, \quad \left\{ \begin{bmatrix} 1 \\ 1 \\ 1 \end{bmatrix} \right\}, \quad \text{and} \quad \left\{ \begin{bmatrix} 0 \\ 1 \\ 0 \end{bmatrix} \right\}.$$

29. The standard matrix of T is

$$\begin{bmatrix} -1 & 6 \\ -8 & 13 \end{bmatrix}.$$

The eigenvalues of T are the roots of its characteristic polynomial, which are 5 and 7. The eigenvectors of T are the same as the eigenvectors of A; so we must find bases for the null spaces of $A - 5I_2$ and $A - 7I_2$. As in Exercise 5, we obtain the following bases for the eigenspaces corresponding to 5 and 7:

$$\left\{ \begin{bmatrix} 1 \\ 1 \end{bmatrix} \right\} \quad \text{and} \quad \left\{ \begin{bmatrix} 3 \\ 4 \end{bmatrix} \right\}.$$

33. The eigenvalues of T are 1 and 2 (with multiplicity 2), and the standard matrix of T is

$$\begin{bmatrix} 3 & 2 & -2 \\ 2 & 6 & -4 \\ 3 & 6 & -4 \end{bmatrix}.$$

As in Exercise 29, we find the following bases for the eigenspaces:

$$\left\{ \begin{bmatrix} 1 \\ 2 \\ 3 \end{bmatrix} \right\} \quad \text{and} \quad \left\{ \begin{bmatrix} -2 \\ 1 \\ 0 \end{bmatrix}, \begin{bmatrix} 2 \\ 0 \\ 1 \end{bmatrix} \right\}.$$

37. The standard matrix of T is

$$\begin{bmatrix} 7 & -10 & 0 \\ 5 & -8 & 0 \\ -1 & 1 & 2 \end{bmatrix},$$

and its characteristic polynomial is $-t^3 + t^2 + 8t - 12 = -(t+3)(t-2)^2$. So the eigenvalues of T are -3 and 2 (with multiplicity 2). As in Exercise 29, we find the following bases for the eigenspaces:

$$\left\{ \begin{bmatrix} 1 \\ 1 \\ 0 \end{bmatrix} \right\} \quad \text{and} \quad \left\{ \begin{bmatrix} 0 \\ 0 \\ 1 \end{bmatrix} \right\}.$$

41. The characteristic polynomial of the given matrix is $t^2 - 3t + 10$, which has no real roots. Hence the given matrix has no real eigenvalues.

45. Suppose that A is an upper triangular or lower triangular $n \times n$ matrix. Then $A - tI_n$ is also an upper triangular or lower triangular matrix, and so the characteristic polynomial of A is

$$\det (A - tI_n) = (a_{11} - t)(a_{22} - t) \cdots (a_{nn} - t)$$

by Theorem 3.2. Thus λ is an eigenvalue of A with multiplicity k if and only if exactly k of the diagonal entries of A equal λ.

49. (a) The characteristic polynomial of A^T is

$$t^2 - 13t + 42 = (t-6)(t-7).$$

(b) The characteristic polynomial of B^T equals that of B since

$$\det (B^T - tI_n) = \det (B - tI_n)^T = \det (B - tI_n).$$

(c) Because the characteristic polynomials of B and B^T are equal, the eigenvalues of B and B^T are the same.

(d) There is no relationship between the eigenvectors of an arbitrary matrix B and those of B^T.

53. The characteristic polynomial of the given matrix is

$$-t^3 + \frac{23}{15}t^2 - \frac{127}{720}t + \frac{1}{2160}.$$

57. (a) The characteristic polynomial of A is

$$t^2 - 2.5t - 1.5 = (t+0.5)(t-3);$$

so the eigenvalues of A are -0.5 and 3. Corresponding eigenvectors are

$$\begin{bmatrix} 1 \\ 2 \end{bmatrix} \quad \text{and} \quad \begin{bmatrix} 1 \\ 1 \end{bmatrix}.$$

(b)   The matrix A is invertible because its rank is 2; in fact,

$$A^{-1} = \frac{1}{3} \begin{bmatrix} 8 & -7 \\ 14 & -13 \end{bmatrix}.$$

The characteristic polynomial of A^{-1} is

$$t^2 + \frac{5}{3}t - \frac{2}{3} = \frac{1}{3}(t+2)(3t-1),$$

and so the corresponding eigenvalues of A^{-1} are -2 and $\frac{1}{3}$. Corresponding eigenvectors are

$$\begin{bmatrix} 1 \\ 2 \end{bmatrix} \quad \text{and} \quad \begin{bmatrix} 1 \\ 1 \end{bmatrix}.$$

(c)   The eigenvalues of an invertible matrix A are the reciprocals of the eigenvalues of its inverse, and an eigenvector of A corresponding to eigenvalue λ is also an eigenvector of A^{-1} corresponding to the eigenvalue $\frac{1}{\lambda}$.

(d)   The eigenvalues of the given matrix are -2, 4, and 5 with corresponding eigenvectors

$$\mathbf{v}_1 = \begin{bmatrix} 0 \\ 1 \\ 1 \end{bmatrix}, \quad \mathbf{v}_2 = \begin{bmatrix} -2 \\ 3 \\ 2 \end{bmatrix}, \quad \text{and} \quad \mathbf{v}_3 = \begin{bmatrix} -1 \\ 2 \\ 1 \end{bmatrix}.$$

The inverse of the given matrix is

$$\begin{bmatrix} .30 & .10 & -.10 \\ -.85 & .05 & -.55 \\ -.80 & -.10 & -.40 \end{bmatrix};$$

its eigenvalues are $-.50$, .25, and .20. Corresponding eigenvectors are $\mathbf{v}_1$, $\mathbf{v}_2$, and $\mathbf{v}_3$.

(e)   Let $\mathbf{v}$ be an eigenvector of an invertible matrix A that corresponds to eigenvalue λ. Then

$$A\mathbf{v} = \lambda\mathbf{v}.$$

Since A is invertible and $\mathbf{v} \neq \mathbf{0}$, we have $A\mathbf{v} \neq \mathbf{0}$. Hence $\lambda \neq 0$. So

$$\mathbf{v} = A^{-1}(\lambda\mathbf{v})$$
$$\mathbf{v} = \lambda(A^{-1}\mathbf{v})$$
$$\frac{1}{\lambda}\mathbf{v} = A^{-1}\mathbf{v}.$$

Thus $\mathbf{v}$ is an eigenvector of A^{-1} corresponding to eigenvalue $\frac{1}{\lambda}$.

5.3 DIAGONALIZATION OF MATRICES

1. **(a)** False, see Example 1.
   (b) True
   (c) True
   (d) True
   (e) False, the eigenvalues of A may occur in any sequence as the diagonal entries of D.
   (f) False, if an $n \times n$ matrix has n *linearly independent* eigenvectors, then it is diagonalizable.
   (g) False, I_n is diagonalizable and has only one eigenvalue.
   (h) True
   (i) False, see Example 1.
   (j) False, for A to be diagonalizable, its characteristic polynomial must also factor as a product of linear factors.
   (k) True
   (l) False, the dimension of the eigenspace corresponding to λ is the *nullity* of $A - \lambda I_n$.

5. **(a)** In order for the matrix to be diagonalizable, the dimension of the eigenspace corresponding to eigenvalue -3 must be **4**.
   (b) If the dimension of the eigenspace corresponding to eigenvalue -3 is **1, 2, or 3**, then the matrix is not diagonalizable.

7. **(a)** Since $\dim W_1 = 2$, the multiplicity of eigenvalue 4 must be at least 2 by Theorem 5.1. Likewise, the multiplicity of eigenvalue 8 must be at least 2, and the multiplicity of eigenvalue 5 must be at least 1. However, A is a 5×5 matrix, and so the sum of the multiplicities of its eigenvalues must be exactly 5. Thus the multiplicities of eigenvalues 4, 5, and 8 must be 2, 1, and 2, respectively, so that the characteristic polynomial of A is $-(t-4)^2(t-5)(t-8)^2$.
   (b) Since A is diagonalizable, the sum of the dimensions of its eigenspaces must be 5 by the test for a diagonalizable matrix. But the given information allows either $\dim W_1 = 2$ or $\dim W_3 = 2$, and so the characteristic polynomial of A cannot be determined from this information.
   (c) As in (b), the sum of the dimensions of the eigenspaces of A must be 5. Hence $\dim W_3 = 2$, so that the characteristic polynomial of A is $-(t-4)(t-5)^2(t-8)^2$.

9. The eigenvalues of A are 4 and 5. Eigenvectors corresponding to eigenvalue 4 are solutions to $(A - 4I_2)\mathbf{x} = \mathbf{0}$. Since the reduced row echelon form of $A - 4I_2$ is

$$\begin{bmatrix} 1 & 2 \\ 0 & 0 \end{bmatrix},$$

these solutions have the form

$$\begin{bmatrix} x_1 \\ x_2 \end{bmatrix} = x_2 \begin{bmatrix} -2 \\ 1 \end{bmatrix}.$$

Hence

$$\left\{ \begin{bmatrix} -2 \\ 1 \end{bmatrix} \right\}$$

is a basis for the eigenspace of A corresponding to eigenvalue 4. Likewise, the reduced row echelon form of $A - 5I_2$ is

$$\begin{bmatrix} 1 & 3 \\ 0 & 0 \end{bmatrix},$$

and so

$$\left\{ \begin{bmatrix} -3 \\ 1 \end{bmatrix} \right\}$$

is a basis for the eigenspace of A corresponding to eigenvalue 5. Let P be the matrix whose columns are the vectors in the bases for the eigenspaces, and let D be the diagonal matrix whose diagonal entries are the corresponding eigenvalues:

$$P = \begin{bmatrix} -2 & -3 \\ 1 & 1 \end{bmatrix} \quad \text{and} \quad D = \begin{bmatrix} 4 & 0 \\ 0 & 5 \end{bmatrix}.$$

Then $A = PDP^{-1}$.

13. Proceed as in Exercise 9. The eigenvalues of A are -5, 2, and 3, and bases for the corresponding eigenspaces are

$$\left\{ \begin{bmatrix} 0 \\ 1 \\ 1 \end{bmatrix} \right\}, \quad \left\{ \begin{bmatrix} -2 \\ 3 \\ 2 \end{bmatrix} \right\}, \quad \text{and} \quad \left\{ \begin{bmatrix} -1 \\ 1 \\ 1 \end{bmatrix} \right\}.$$

Hence we may take

$$P = \begin{bmatrix} 0 & -2 & -1 \\ 1 & 3 & 1 \\ 1 & 2 & 1 \end{bmatrix} \quad \text{and} \quad D = \begin{bmatrix} -5 & 0 & 0 \\ 0 & 2 & 0 \\ 0 & 0 & 3 \end{bmatrix}.$$

17. Proceed as in Exercise 9. The eigenvalues of A are 3 and 5, and bases for the corresponding eigenspaces are

$$\left\{ \begin{bmatrix} -1 \\ 1 \\ 0 \end{bmatrix}, \begin{bmatrix} 1 \\ 0 \\ 1 \end{bmatrix} \right\} \quad \text{and} \quad \left\{ \begin{bmatrix} -1 \\ 4 \\ 2 \end{bmatrix} \right\}.$$

Hence we may take

$$P = \begin{bmatrix} -1 & 1 & -1 \\ 1 & 0 & 4 \\ 0 & 1 & 2 \end{bmatrix} \quad \text{and} \quad D = \begin{bmatrix} 3 & 0 & 0 \\ 0 & 3 & 0 \\ 0 & 0 & 5 \end{bmatrix}.$$

21. The characteristic polynomial of A is $t^2 - 2t + 1 = (t-1)^2$. Since the rank of $A - I_2$ is 1, the eigenspace of A corresponding to eigenvalue 1 is 1-dimensional. Hence A is not diagonalizable because the eigenvalue 1 has multiplicity 2 and its eigenspace is 1-dimensional.

25. The characteristic polynomial of A is

$$-t^3 - 2t^2 + 5t + 6 = -(t+3)(t+1)(t-2).$$

Bases for the corresponding eigenspaces are

$$\left\{ \begin{bmatrix} -1 \\ 1 \\ 0 \end{bmatrix} \right\}, \quad \left\{ \begin{bmatrix} 1 \\ 0 \\ 0 \end{bmatrix} \right\}, \quad \text{and} \quad \left\{ \begin{bmatrix} -1 \\ 1 \\ 5 \end{bmatrix} \right\}.$$

Hence we may take

$$P = \begin{bmatrix} -1 & 1 & -1 \\ 1 & 0 & 1 \\ 0 & 0 & 5 \end{bmatrix} \quad \text{and} \quad D = \begin{bmatrix} -3 & 0 & 0 \\ 0 & -1 & 0 \\ 0 & 0 & 2 \end{bmatrix}.$$

29. We have $A = PDP^{-1}$, where

$$P = \begin{bmatrix} 1 & 2 \\ 1 & 1 \end{bmatrix} \quad \text{and} \quad D = \begin{bmatrix} 4 & 0 \\ 0 & 3 \end{bmatrix}.$$

Thus, as in the example on page 270, we have

$$A^k = PD^kP^{-1} = \begin{bmatrix} 1 & 2 \\ 1 & 1 \end{bmatrix} \begin{bmatrix} 4^k & 0 \\ 0 & 3^k \end{bmatrix} \begin{bmatrix} -1 & 2 \\ 1 & -1 \end{bmatrix}$$

$$= \begin{bmatrix} 4^k & 2 \cdot 3^k \\ 4^k & 3^k \end{bmatrix} \begin{bmatrix} -1 & 2 \\ 1 & -1 \end{bmatrix} = \begin{bmatrix} 2 \cdot 3^k - 4^k & 2 \cdot 4^k - 2 \cdot 3^k \\ 3^k - 4^k & 2 \cdot 4^k - 3^k \end{bmatrix}.$$

33. We have $A = PDP^{-1}$, where

$$P = \begin{bmatrix} -1 & 0 & -2 \\ 1 & 0 & 1 \\ 0 & 1 & 0 \end{bmatrix} \quad \text{and} \quad D = \begin{bmatrix} 5 & 0 & 0 \\ 0 & 5 & 0 \\ 0 & 0 & 1 \end{bmatrix}.$$

Then, as in the example on page 270, we have

$$A^k = PD^kP^{-1} = \begin{bmatrix} -1 & 0 & -2 \\ 1 & 0 & 1 \\ 0 & 1 & 0 \end{bmatrix} \begin{bmatrix} 5^k & 0 & 0 \\ 0 & 5^k & 0 \\ 0 & 0 & 1 \end{bmatrix} \begin{bmatrix} 1 & 2 & 0 \\ 0 & 0 & 1 \\ -1 & -1 & 0 \end{bmatrix}$$

$$= \begin{bmatrix} -5^k & 0 & -2 \\ 5^k & 0 & 1 \\ 0 & 5^k & 0 \end{bmatrix} \begin{bmatrix} 1 & 2 & 0 \\ 0 & 0 & 1 \\ -1 & -1 & 0 \end{bmatrix} = \begin{bmatrix} 2 - 5^k & 2 - 2 \cdot 5^k & 0 \\ 5^k - 1 & 2 \cdot 5^k - 1 & 0 \\ 0 & 0 & 5^k \end{bmatrix}.$$

37. Since the characteristic polynomial of the given matrix does not factor into a product of linear factors, this matrix is not diagonalizable for any scalar c.

41. It follows from the second boxed statement on page 277 that the given matrix is diagonalizable if $c \neq -2$ and $c \neq -1$. Thus we must check only the values of -2 and -1. For $c = -2$, we see that -2 is an eigenvalue of multiplicity 2, but the reduced row echelon form of $A + 2I_3$, which is

$$\begin{bmatrix} 1 & 0 & 0 \\ 0 & 0 & 1 \\ 0 & 0 & 0 \end{bmatrix},$$

has rank 2. Hence A is not diagonalizable if $c = -2$. Likewise, for $c = -1$, the eigenvalue -1 has multiplicity 2, but the reduced row echelon form of $A + I_3$, which is

$$\begin{bmatrix} 1 & 0 & 0 \\ 0 & 0 & 1 \\ 0 & 0 & 0 \end{bmatrix},$$

has rank 2. Thus A is also not diagonalizable if $c = -1$.

45. The desired matrix A satisfies $A = PDP^{-1}$, where

$$P = \begin{bmatrix} 1 & 1 \\ 1 & 3 \end{bmatrix} \quad \text{and} \quad D = \begin{bmatrix} -3 & 0 \\ 0 & 5 \end{bmatrix}.$$

(Here the columns of P are the given eigenvectors of A, and the diagonal entries of D are the corresponding eigenvalues.) Thus

$$A = \begin{bmatrix} -7 & 4 \\ -12 & 9 \end{bmatrix}.$$

49. Let D be a diagonal matrix. Then $D = I_n D I_n^{-1}$, and so D is diagonalizable because I_n is invertible.

53. Let A be a diagonalizable $n \times n$ matrix. Then $A = PDP^{-1}$, for some diagonal matrix D and invertible matrix P. Now D^2 is a diagonal matrix (see page 90), and

$$A^2 = (PDP^{-1})(PDP^{-1}) = (PD^2P^{-1}).$$

Hence A^2 is diagonalizable.

57. Let A be a diagonalizable $n \times n$ matrix that is nilpotent. Then by Exercise 44 of Section 5.1, the only eigenvalue of A is 0. Hence by Exercise 50(a), we have $A = 0I_n = O$.

61. The characteristic polynomial of the given matrix is

$$t^4 + 6t^3 + 13t^2 + 12t + 4 = (t+2)^2(t+1)^2.$$

Since the parametric representation of the general solution to $(A + 2I_4)\mathbf{x} = \mathbf{0}$ is

$$\begin{bmatrix} x_1 \\ x_2 \\ x_3 \\ x_4 \end{bmatrix} = x_3 \begin{bmatrix} -\frac{8}{3} \\ -\frac{1}{3} \\ 1 \\ 0 \end{bmatrix} + x_4 \begin{bmatrix} -\frac{1}{3} \\ -\frac{2}{3} \\ 0 \\ 1 \end{bmatrix},$$

and the parametric representation of the general solution to $(A + I_4)\mathbf{x} = \mathbf{0}$ is

$$\begin{bmatrix} x_1 \\ x_2 \\ x_3 \\ x_4 \end{bmatrix} = x_3 \begin{bmatrix} -\frac{3}{2} \\ -\frac{1}{2} \\ 1 \\ 0 \end{bmatrix} + x_3 \begin{bmatrix} -\frac{1}{2} \\ -\frac{1}{2} \\ 0 \\ 1 \end{bmatrix},$$

the sets

$$\left\{ \begin{bmatrix} -8 \\ -1 \\ 3 \\ 0 \end{bmatrix}, \begin{bmatrix} -1 \\ -2 \\ 0 \\ 3 \end{bmatrix} \right\} \quad \text{and} \quad \left\{ \begin{bmatrix} -3 \\ -1 \\ 2 \\ 0 \end{bmatrix}, \begin{bmatrix} -1 \\ -1 \\ 0 \\ 2 \end{bmatrix} \right\}$$

are bases for the eigenspaces of A corresponding to eigenvalues -2 and -1, respectively. Thus $A = PDP^{-1}$, where

$$P = \begin{bmatrix} -8 & -1 & -3 & -1 \\ -1 & -2 & -1 & -1 \\ 3 & 0 & 2 & 0 \\ 0 & 3 & 0 & 2 \end{bmatrix} \quad \text{and} \quad D = \begin{bmatrix} -2 & 0 & 0 & 0 \\ 0 & -2 & 0 & 0 \\ 0 & 0 & -1 & 0 \\ 0 & 0 & 0 & -1 \end{bmatrix}.$$

5.4 DIAGONALIZATION OF LINEAR OPERATORS

1. (a) False, its standard matrix is diagonalizable, that is, *similar* to a diagonal matrix.
   (b) False, the linear operator on $\mathcal{R}^2$ that rotates a vector by 90° is not diagonalizable.
   (c) True
   (d) False, $\mathcal{B}$ can be any basis for $\mathcal{R}^n$ consisting of eigenvectors of T.
   (e) False, the eigenvalues of T may occur in any sequence as the diagonal entries of D.
   (f) False, any vector perpendicular to W belongs to the null space of the orthogonal projection operator.
   (g) False, the range of the operator is W, not $\mathcal{R}^3$.

5. The standard matrix of T is

$$A = \begin{bmatrix} -3 & 5 & -5 \\ 2 & -3 & 2 \\ 2 & -5 & 4 \end{bmatrix}.$$

If B is the matrix whose columns are the vectors in $\mathcal{B}$, then

$$[T]_\mathcal{B} = B^{-1}AB = \begin{bmatrix} 2 & 0 & 0 \\ 0 & -1 & 0 \\ 0 & 0 & -3 \end{bmatrix}.$$

Since $[T]_\mathcal{B}$ is a diagonal matrix, the basis $\mathcal{B}$ consists of eigenvectors of T.

9. The standard matrix of T is

$$A = \begin{bmatrix} 7 & -5 \\ 10 & -8 \end{bmatrix}.$$

A basis for the eigenspace of T corresponding to eigenvalue -3 is obtained by solving $(A + 3I_2)\mathbf{x} = \mathbf{0}$, and a basis for the eigenspace of T corresponding to eigenvalue 2 is obtained by solving $(A - 2I_2)\mathbf{x} = \mathbf{0}$. The resulting bases are

$$\mathcal{B}_1 = \left\{ \begin{bmatrix} 1 \\ 2 \end{bmatrix} \right\} \quad \text{and} \quad \mathcal{B}_1 = \left\{ \begin{bmatrix} 1 \\ 1 \end{bmatrix} \right\}.$$

Then $\mathcal{B}_1 \cup \mathcal{B}_2$ is a basis for $\mathcal{R}^2$ consisting of eigenvectors of T.

13. The standard matrix of T is

$$A = \begin{bmatrix} -1 & -1 & 0 \\ 0 & -1 & 0 \\ 1 & 1 & 0 \end{bmatrix}.$$

Since the reduced row echelon form of $A + I_3$ is

$$\begin{bmatrix} 1 & 0 & 1 \\ 0 & 1 & 0 \\ 0 & 0 & 0 \end{bmatrix},$$

the dimension of the eigenspace of T corresponding to eigenvalue -1 is

$$3 - \text{rank}\,(A + I_3) = 1.$$

But the multiplicity of the eigenvalue -1 is 2, so that T is not diagonalizable. That is, there is no basis for $\mathcal{R}^3$ consisting of eigenvectors of T.

17. The standard matrix of T is

$$\begin{bmatrix} -7 & -4 & 4 & -4 \\ 0 & 1 & 0 & 0 \\ -8 & -4 & 5 & -4 \\ 0 & 0 & 0 & 1 \end{bmatrix}.$$

As in Exercise 9, we find that

$$\mathcal{B}_1 = \left\{ \begin{bmatrix} 1 \\ 0 \\ 1 \\ 0 \end{bmatrix} \right\} \quad \text{and} \quad \mathcal{B}_2 = \left\{ \begin{bmatrix} -1 \\ 2 \\ 0 \\ 0 \end{bmatrix}, \begin{bmatrix} 1 \\ 0 \\ 2 \\ 0 \end{bmatrix}, \begin{bmatrix} -1 \\ 0 \\ 0 \\ 2 \end{bmatrix} \right\}$$

are bases for the eigenspaces of T corresponding to eigenvalues -3 and 1, respectively. Then $\mathcal{B}_1 \cup \mathcal{B}_2$ is a basis for $\mathcal{R}^4$ consisting of eigenvectors of T.

21. The standard matrix for T is

$$\begin{bmatrix} -2 & 3 \\ 4 & -3 \end{bmatrix},$$

and its characteristic polynomial is $t^2 + 5t - 6 = (t + 6)(t - 1)$. As in Exercise 9, we find that

$$\mathcal{B}_1 = \left\{ \begin{bmatrix} -3 \\ 4 \end{bmatrix} \right\} \quad \text{and} \quad \mathcal{B}_1 = \left\{ \begin{bmatrix} 1 \\ 1 \end{bmatrix} \right\}$$

are bases for the eigenspaces of T corresponding to the eigenvalues -6 and 1, respectively. Thus $\mathcal{B} = \mathcal{B}_1 \cup \mathcal{B}_2$ is a basis for $\mathcal{R}^2$ consisting of eigenvectors of T, and so

$$[T]_\mathcal{B} = \begin{bmatrix} -6 & 0 \\ 0 & 1 \end{bmatrix}$$

is a diagonal matrix.

25. The standard matrix of T is

$$\begin{bmatrix} 1 & 0 & 0 \\ -1 & 1 & -1 \\ 0 & 0 & 1 \end{bmatrix},$$

and its characteristic polynomial is $-t^3 + 3t^2 - 3t + 1 = -(t - 1)^3$. Since the reduced row echelon form of $A - I_3$ is

$$\begin{bmatrix} 1 & 0 & 1 \\ 0 & 0 & 0 \\ 0 & 0 & 0 \end{bmatrix},$$

the dimension of the eigenspace of T corresponding to eigenvalue 1 is

$$3 - \text{rank}\,(A - I_3) = 2.$$

Because this dimension does not equal the multiplicity of the eigenvalue 1, T is not diagonalizable. That is, there is no basis B for $\mathcal{R}^3$ such that $[T]_B$ is a diagonal matrix.

29. The standard matrix of T is

$$A = \begin{bmatrix} c & 0 & 0 \\ -1 & -3 & -1 \\ -8 & 1 & -5 \end{bmatrix},$$

and so

$$A + 4I_3 = \begin{bmatrix} c+4 & 0 & 0 \\ -1 & 1 & -1 \\ -8 & 1 & -1 \end{bmatrix}.$$

Because the last two rows of $A + 4I_3$ are linearly independent, the rank of $A + 4I_3$ is at least 2. Hence the dimension of the eigenspace of T corresponding to eigenvalue -4 is 1. Since this dimension does not equal the multiplicity of the eigenvalue -4, T is not diagonalizable for any scalar c.

31. The standard matrix of T is

$$A = \begin{bmatrix} c & 0 & 0 \\ 2 & -3 & 2 \\ -3 & 0 & 1 \end{bmatrix}.$$

The characteristic polynomial of A is the same as that of T, namely, $-(t-c)(t+3)(t+1)$. If $c \neq -3$ and $c \neq -1$, then the second boxed statement on page 277 shows that A (and hence T) is diagonalizable.

If $c = -3$, then -3 is an eigenvalue of A (and T) with multiplicity 2. But the reduced echelon form of $A + 3I_3$ is

$$\begin{bmatrix} 1 & 0 & 0 \\ 0 & 0 & 1 \\ 0 & 0 & 0 \end{bmatrix},$$

so that the eigenspace of A (and T) corresponding to eigenvalue -3 has dimension $3 - 2 = 1$. Since the second condition in the test for a diagonalizable linear operator fails, T is not diagonalizable if $c = -3$.

If $c = -1$, then -1 is an eigenvalue of A (and T) with multiplicity 2. But the reduced row echelon form of $A + I_3$ is

$$\begin{bmatrix} 1 & 0 & 0 \\ 0 & 1 & -1 \\ 0 & 0 & 0 \end{bmatrix},$$

so that the eigenspace of A (and T) corresponding to eigenvalue -1 has dimension $3 - 2 = 1$. Since the second condition in the test for a diagonalizable linear operator fails, T is also not diagonalizable if $c = -1$.

33. Since the characteristic polynomial of T does not factor into linear factors, T is not diagonalizable for any scalar c.

37. The parametric representation of the general solution to $x + y + z = 0$ is

$$\begin{bmatrix} x \\ y \\ z \end{bmatrix} = y \begin{bmatrix} -1 \\ 1 \\ 0 \end{bmatrix} + z \begin{bmatrix} -1 \\ 0 \\ 1 \end{bmatrix}.$$

Hence

$$\left\{ \begin{bmatrix} -1 \\ 1 \\ 0 \end{bmatrix}, \begin{bmatrix} -1 \\ 0 \\ 1 \end{bmatrix} \right\}$$

is a basis for W, the eigenspace of U corresponding to eigenvalue 1. As on page 287, the vector

$$\begin{bmatrix} 1 \\ 1 \\ 1 \end{bmatrix}$$

whose components are the coefficients of the equation $x + y + z = 0$, is normal to W, and so is an eigenvector of U corresponding to eigenvalue 0. Thus

$$\mathcal{B} = \left\{ \begin{bmatrix} -1 \\ 1 \\ 0 \end{bmatrix}, \begin{bmatrix} -1 \\ 0 \\ 1 \end{bmatrix}, \begin{bmatrix} 1 \\ 1 \\ 1 \end{bmatrix} \right\}$$

is a basis for $\mathcal{R}^3$ consisting of eigenvectors of U. Hence

$$[U]_\mathcal{B} = \begin{bmatrix} 1 & 0 & 0 \\ 0 & 1 & 0 \\ 0 & 0 & 0 \end{bmatrix}.$$

Let B be the matrix whose columns are the vectors in $\mathcal{B}$. Then the standard matrix of U is

$$A = B[U]_\mathcal{B} B^{-1} = \frac{1}{3} \begin{bmatrix} 2 & -1 & -1 \\ -1 & 2 & -1 \\ -1 & -1 & 2 \end{bmatrix}.$$

Therefore

$$U\left(\begin{bmatrix} x_1 \\ x_2 \\ x_3 \end{bmatrix} \right) = A \begin{bmatrix} x_1 \\ x_2 \\ x_3 \end{bmatrix} = \frac{1}{3} \begin{bmatrix} 2x_1 - x_2 - x_3 \\ -x_1 + 2x_2 - x_3 \\ -x_1 - x_2 + 2x_3 \end{bmatrix}.$$

41. As in Exercise 37,

$$B = \left\{ \begin{bmatrix} -1 \\ 1 \\ 0 \end{bmatrix}, \begin{bmatrix} -1 \\ 0 \\ 2 \end{bmatrix}, \begin{bmatrix} 2 \\ 2 \\ 1 \end{bmatrix} \right\}$$

is a basis for $\mathcal{R}^3$ consisting of eigenvectors of U. Thus $[U]_B = [\mathbf{e}_1 \ \mathbf{e}_2 \ \mathbf{0}]$. So if B is the matrix whose columns are the vectors in B and A is the standard matrix of U, then

$$A = B[U]_B B^{-1} = \frac{1}{9} \begin{bmatrix} 5 & -4 & -2 \\ -4 & 5 & -2 \\ -2 & -2 & 8 \end{bmatrix}.$$

Therefore

$$U\left(\begin{bmatrix} x_1 \\ x_2 \\ x_3 \end{bmatrix} \right) = A \begin{bmatrix} x_1 \\ x_2 \\ x_3 \end{bmatrix} = \frac{1}{9} \begin{bmatrix} 5x_1 - 4x_2 - 2x_3 \\ -4x_1 + 5x_2 - 2x_3 \\ -2x_1 - 2x_2 + 8x_3 \end{bmatrix}.$$

45. As in Exercise 37,

$$B = \left\{ \begin{bmatrix} -2 \\ 1 \\ 0 \end{bmatrix}, \begin{bmatrix} 1 \\ 0 \\ 1 \end{bmatrix}, \begin{bmatrix} 1 \\ 2 \\ -1 \end{bmatrix} \right\}$$

is a basis for $\mathcal{R}^3$ consisting of eigenvectors of T. Since the third vector in B is an eigenvector of T corresponding to eigenvalue -1, we have

$$[T]_B = \begin{bmatrix} 1 & 0 & 0 \\ 0 & 1 & 0 \\ 0 & 0 & -1 \end{bmatrix}.$$

Let B be the matrix whose columns are the vectors in B and A be the standard matrix of T. Then

$$A = B[T]_B B^{-1} = \frac{1}{3} \begin{bmatrix} 2 & -2 & 1 \\ -2 & -1 & 2 \\ 1 & 2 & 2 \end{bmatrix}.$$

Therefore

$$T\left(\begin{bmatrix} x_1 \\ x_2 \\ x_3 \end{bmatrix} \right) = A \begin{bmatrix} x_1 \\ x_2 \\ x_3 \end{bmatrix} = \frac{1}{3} \begin{bmatrix} 2x_1 - 2x_2 + x_3 \\ -2x_1 - x_2 + 2x_3 \\ x_1 + 2x_2 + 2x_3 \end{bmatrix}.$$

49. Let $\mathbf{z}_1$ and $\mathbf{z}_2$ be in $\mathcal{R}^3$. Because $\{\mathbf{u}, \mathbf{v}, \mathbf{w}\}$ is a basis for $\mathcal{R}^3$, there exist unique scalars a_i, b_i, c_i such that $\mathbf{z}_i = a_i\mathbf{u} + b_i\mathbf{v} + c_i\mathbf{w}$ for $i = 1, 2$. Now

$$T(\mathbf{z}_1 + \mathbf{z}_2) = T((a_1 + a_2)\mathbf{u} + (b_1 + b_2)\mathbf{v} + (c_1 + c_2)\mathbf{w})$$
$$= (a_1 + a_2)\mathbf{u} + (b_1 + b_2)\mathbf{v},$$

and

$$T(\mathbf{z}_1) + T(\mathbf{z}_2) = T(a_1\mathbf{u} + b_1\mathbf{v} + c_1\mathbf{w}) + T(a_2\mathbf{u} + b_2\mathbf{v} + c_2\mathbf{w})$$
$$= (a_1\mathbf{u} + b_1\mathbf{v}) + (a_2\mathbf{u} + b_2\mathbf{v})$$
$$= (a_1 + a_2)\mathbf{u} + (b_1 + b_2)\mathbf{v}.$$

So T preserves vector addition. In addition, for any scalar d,

$$T(d\mathbf{z}_1) = T((da_1)\mathbf{u} + (db_1)\mathbf{v} + (dc_1)\mathbf{w})$$
$$= (da_1)\mathbf{u} + (db_1)\mathbf{v}$$
$$= d(a_1\mathbf{u} + b_1\mathbf{v})$$
$$= dT(\mathbf{z}_1).$$

Hence T also preserves scalar multiplication.

53. Yes, cT is diagonalizable for any scalar c. Let $\mathbf{v}$ be an eigenvector of T with corresponding eigenvalue λ. Then

$$(cT)(\mathbf{v}) = c(T(\mathbf{v})) = c(\lambda\mathbf{v}) = (c\lambda)\mathbf{v},$$

and so $\mathbf{v}$ is also an eigenvector of cT (with corresponding eigenvalue $c\lambda$). Thus if $\mathcal{B}$ is a basis for $\mathcal{R}^n$ consisting of eigenvectors of T, then $\mathcal{B}$ is also a basis for $\mathcal{R}^n$ consisting of eigenvectors of cT.

57. Let U be a diagonalizable linear operator on $\mathcal{R}^n$ having only nonnegative eigenvalues, and let $\mathcal{B}$ be a basis for $\mathcal{R}^n$ consisting of eigenvectors of U. If C is the standard matrix of U and B is the matrix whose columns are the vectors in $\mathcal{B}$, then $[U]_\mathcal{B} = B^{-1}CB$. Let A be the diagonal matrix whose entries are the square roots of the entries of the diagonal matrix $[U]_\mathcal{B}$. Then $A^2 = [U]_\mathcal{B}$, and so

$$C = B[U]_\mathcal{B}B^{-1} = BA^2B^{-1} = (BAB^{-1})(BAB^{-1}) = (BAB^{-1})^2.$$

Let T be the matrix transformation induced by BAB^{-1}.

5.5   APPLICATIONS OF EIGENVALUES

1. (a) False, the *column* sums of the transition matrix of a Markov chain are all 1.
    (b) False, see the matrix A on page 292.
    (c) True
    (d) False, consider $A = \begin{bmatrix} 0 & 1 \\ 1 & 0 \end{bmatrix}$ and $\mathbf{p} = \begin{bmatrix} .8 \\ .2 \end{bmatrix}$.
    (e) True
    (f) False, the general solution to $y' = ky$ is $y = ce^{kt}$.
    (g) False, the change of variable $\mathbf{y} = P\mathbf{z}$ transforms $\mathbf{y}' = A\mathbf{y}$ into $\mathbf{z}' = D\mathbf{z}$.
    (h) True

5. Since there are no zero entries in

$$A^3 = \begin{bmatrix} .475 & .35 & .385 \\ .275 & .35 & .265 \\ .250 & .30 & .350 \end{bmatrix},$$

   A is a regular transition matrix.

9. A steady-state vector is a probability vector that is also an eigenvector corresponding to eigenvalue 1. We begin by finding the eigenvectors corresponding to eigenvalue 1. Since the reduced row echelon form of $A - I_3$ is

$$\begin{bmatrix} 1 & 0 & -.5 \\ 0 & 1 & -.5 \\ 0 & 0 & 0 \end{bmatrix},$$

   a basis for the eigenspace corresponding to eigenvalue 1 is

$$\left\{ \begin{bmatrix} 1 \\ 1 \\ 2 \end{bmatrix} \right\}.$$

   Thus the eigenvectors corresponding to eigenvalue 1 have the form

$$c \begin{bmatrix} 1 \\ 1 \\ 2 \end{bmatrix} = \begin{bmatrix} c \\ c \\ 2c \end{bmatrix}$$

   for some nonzero scalar c. We seek a vector of this form that is also a probability vector, that is, such that

$$c + c + 2c = 1.$$

   So $c = .25$, and the steady-state vector is

$$\begin{bmatrix} .25 \\ .25 \\ .50 \end{bmatrix}.$$

13. (a) The two states of this Markov chain are buying a root beer float (F) and buying a chocolate sundae (S). A transition matrix for this Markov chain is

$$\text{Next visit} \quad \begin{array}{c} \\ F \\ S \end{array} \overset{\begin{array}{cc} \text{Last visit} \\ F \quad\quad S \end{array}}{\begin{bmatrix} .25 & .5 \\ .75 & .5 \end{bmatrix}} = A.$$

Note that the $(1,2)$-entry and the $(2,1)$-entry of A can be determined from the condition that each column sum in A must be 1.

(b) If Alison bought a sundae on her next-to-last visit, we can take

$$\mathbf{p} = \begin{bmatrix} 0 \\ 1 \end{bmatrix}.$$

Then the probabilities of each purchase on her last visit are

$$A\mathbf{p} = \begin{bmatrix} .5 \\ .5 \end{bmatrix},$$

and the probabilities of each purchase on her next visit are

$$A^2\mathbf{p} = A(A\mathbf{p}) = \begin{bmatrix} .375 \\ .625 \end{bmatrix}.$$

Thus the probability that she will buy a float on her next visit is .375.

(c) Over the long run, the proportion of purchases of each kind is given by the steady-state vector for A. As in Exercise 9, we first find a basis for the eigenspace corresponding to eigenvalue 1, which is

$$\left\{ \begin{bmatrix} 2 \\ 3 \end{bmatrix} \right\}.$$

The vector in this eigenspace that is also a probability vector is

$$.2 \begin{bmatrix} 2 \\ 3 \end{bmatrix} = \begin{bmatrix} .4 \\ .6 \end{bmatrix}.$$

Hence, over the long run, Alison buys a sundae on 60% of her trips to the ice cream store.

17. (a) This Markov chain has two states: wet (W) and dry (D). A transition matrix is

$$\text{Next day} \quad \begin{array}{c} \\ W \\ D \end{array} \overset{\begin{array}{cc} \text{Current day} \\ W \quad\quad\quad D \end{array}}{\begin{bmatrix} \frac{117}{195} & \frac{80}{615} \\ \frac{78}{195} & \frac{535}{615} \end{bmatrix}} = A.$$

(b)   The probability that the following day will be dry if the current day is dry is the $(2,2)$-entry of A, which is $\frac{535}{615} = \frac{107}{123}$.

(c)   A dry day corresponds to the probability vector

$$\mathbf{p} = \begin{bmatrix} 0 \\ 1 \end{bmatrix}.$$

If a Tuesday in November was dry, then the probabilities of each type of weather on the following Thursday are given by

$$A^2\mathbf{p} = A(A\mathbf{p}) = A\begin{bmatrix} \frac{16}{123} \\ \frac{107}{123} \end{bmatrix} = \begin{bmatrix} \frac{14{,}464}{75{,}645} \\ \frac{61{,}181}{75{,}645} \end{bmatrix} \approx \begin{bmatrix} .191 \\ .809 \end{bmatrix}.$$

So the probability that the following Thursday will be dry is about .809.

(d)   As in (c), the probabilities of each type of weather on Saturday if the previous Wednesday was wet are given by

$$A^3\begin{bmatrix} 1 \\ 0 \end{bmatrix} = \begin{bmatrix} .324 \\ .676 \end{bmatrix}.$$

So the probability that Saturday will be dry if the previous Wednesday was wet is about .676.

(e)   Over the long run, the probability of each type of weather is given by the steady-state vector, which can be computed as in Exercise 9. Since the reduced row echelon form of $A - I_2$ is

$$\begin{bmatrix} 1 & -\frac{40}{123} \\ 0 & 0 \end{bmatrix},$$

the steady state vector is

$$\frac{1}{163}\begin{bmatrix} 40 \\ 123 \end{bmatrix} \approx \begin{bmatrix} .245 \\ .755 \end{bmatrix}.$$

So, over the long run, the probability of a wet day is about .245.

21. (a)   The sum of the entries of each row of A^T is 1. Hence $A^T\mathbf{u} = \mathbf{u}$.

(b)   It follows from (a) that 1 is an eigenvalue of A^T.

(c)   Since 1 is an eigenvalue of A^T, $\det(A^T - I_n) = 0$. Hence

$$\det(A - I_n) = \det(A - I_n)^T = \det(A^T - I_n) = 0.$$

(d)   It follows from (c) that 1 is an eigenvalue of A.

25. (a)   The absolute value of the ith component of $A^T\mathbf{v}$ is

$$|a_{1i}v_1 + a_{2i}v_2 + \cdots + a_{ni}v_n| \le |a_{1i}||v_1| + |a_{2i}||v_2| + \cdots + |a_{ni}||v_n|$$

126

$$\leq |a_{1i}||v_k| + |a_{2i}||v_k| + \cdots + |a_{ni}||v_k|$$
$$= (|a_{1i}| + |a_{2i}| + \cdots + |a_{ni}|)|v_k|$$
$$= |v_k|.$$

(b) Let $\mathbf{v}$ be an eigenvector of A^T corresponding to eigenvalue λ. Then $A^T\mathbf{v} = \lambda\mathbf{v}$. It follows from (a) that the absolute value of the kth component of $A^T\mathbf{v}$ is $|\lambda v_k| \leq |v_k|$. Hence $|\lambda| \cdot |v_k| \leq |v_k|$.

(c) Since $|v_k| \neq 0$, the preceding inequality implies that $|\lambda| \leq 1$.

29. The given system of differential equations can be written in the form $\mathbf{y}' = A\mathbf{y}$, where

$$A = \begin{bmatrix} 2 & 4 \\ -6 & -8 \end{bmatrix}.$$

Now the characteristic polynomial of A is $t^2 + 6t + 8 = (t+4)(t+2)$; so A has eigenvalues of -4 and -2. Bases for the corresponding eigenspaces of A are

$$\left\{ \begin{bmatrix} -2 \\ 3 \end{bmatrix} \right\} \quad \text{and} \quad \left\{ \begin{bmatrix} -1 \\ 1 \end{bmatrix} \right\}.$$

Hence $A = PDP^{-1}$, where

$$P = \begin{bmatrix} -2 & -1 \\ 3 & 1 \end{bmatrix} \quad \text{and} \quad D = \begin{bmatrix} -4 & 0 \\ 0 & -2 \end{bmatrix}.$$

The solution to $\mathbf{z}' = D\mathbf{z}$ is

$$\mathbf{z}_1 = ae^{-4t}$$
$$\mathbf{z}_2 = ae^{-2t}.$$

The algorithm on page 296 gives the solution to the original system to be

$$\begin{bmatrix} y_1 \\ y_2 \end{bmatrix} = \mathbf{y} = P\mathbf{z} = \begin{bmatrix} -2 & -1 \\ 3 & 1 \end{bmatrix} \begin{bmatrix} ae^{-4t} \\ be^{-2t} \end{bmatrix} = \begin{bmatrix} -2ae^{-4t} - be^{-2t} \\ 3ae^{-4t} + be^{-2t} \end{bmatrix}.$$

33. The given system of differential equations can be written in the form $\mathbf{y}' = A\mathbf{y}$, where

$$A = \begin{bmatrix} 1 & 1 \\ 4 & 1 \end{bmatrix}.$$

Since the characteristic polynomial of A is $t^2 - 2t - 3 = (t+1)(t-3)$, A has eigenvalues of -1 and 3. Bases for the corresponding eigenspaces of A are

$$\left\{ \begin{bmatrix} -1 \\ 2 \end{bmatrix} \right\} \quad \text{and} \quad \left\{ \begin{bmatrix} 1 \\ 2 \end{bmatrix} \right\}.$$

Hence $A = PDP^{-1}$, where

$$P = \begin{bmatrix} -1 & 1 \\ 2 & 2 \end{bmatrix} \quad \text{and} \quad D = \begin{bmatrix} -1 & 0 \\ 0 & 3 \end{bmatrix}.$$

The solution of $\mathbf{z}' = D\mathbf{z}$ is

$$z_1 = ae^{-t}$$
$$z_2 = be^{3t}.$$

Thus the general solution to the original system is

$$\begin{bmatrix} y_1 \\ y_2 \end{bmatrix} = \mathbf{y} = P\mathbf{z} = \begin{bmatrix} -1 & 1 \\ 2 & 2 \end{bmatrix} \begin{bmatrix} ae^{-t} \\ be^{3t} \end{bmatrix} = \begin{bmatrix} -ae^{-t} + be^{3t} \\ 2ae^{-t} + 2be^{3t} \end{bmatrix}.$$

Taking $t = 0$, we obtain

$$15 = y_1(0) = -a + b$$

and

$$-10 = y_2(0) = 2a + 2b.$$

Solving this system, we obtain $a = -10$ and $b = 5$. Thus the solution to the original system of differential equations and initial conditions is

$$y_1 = 10e^{-t} + 5e^{3t}$$
$$y_2 = -20e^{-t} + 10e^{3t}.$$

37. The given system of differential equations can be written in the form $\mathbf{y}' = A\mathbf{y}$, where

$$A = \begin{bmatrix} 6 & -5 & -7 \\ 1 & 0 & -1 \\ 3 & -3 & -4 \end{bmatrix}.$$

Since the characteristic polynomial of A is $-t^3 + 2t^2 + t - 2 = -(t+1)(t-1)(t-2)$, A has eigenvalues of -1, 1, and 2. Bases for the corresponding eigenspaces of A are

$$\left\{ \begin{bmatrix} 1 \\ 0 \\ 1 \end{bmatrix} \right\}, \quad \left\{ \begin{bmatrix} 1 \\ 1 \\ 0 \end{bmatrix} \right\}, \quad \text{and} \quad \left\{ \begin{bmatrix} 3 \\ 1 \\ 1 \end{bmatrix} \right\}.$$

Hence $A = PDP^{-1}$, where

$$P = \begin{bmatrix} 1 & 1 & 3 \\ 0 & 1 & 1 \\ 1 & 0 & 1 \end{bmatrix} \quad \text{and} \quad D = \begin{bmatrix} -1 & 0 & 0 \\ 0 & 1 & 0 \\ 0 & 0 & 2 \end{bmatrix}.$$

The solution of $\mathbf{z}' = D\mathbf{z}$ is

$$z_1 = ae^{-t}$$
$$z_2 = be^t$$
$$z_3 = ce^{2t}.$$

Thus the general solution to the original system is

$$\begin{bmatrix} y_1 \\ y_2 \\ y_3 \end{bmatrix} = \mathbf{y} = P\mathbf{z} = \begin{bmatrix} 1 & 1 & 3 \\ 0 & 1 & 1 \\ 1 & 0 & 1 \end{bmatrix} \begin{bmatrix} ae^{-t} \\ be^t \\ ce^{2t} \end{bmatrix} = \begin{bmatrix} ae^{-t} + be^t + 3ce^{2t} \\ be^t + ce^{2t} \\ ae^{-t} + ce^{2t} \end{bmatrix}.$$

Taking $t = 0$, we have

$$0 = y_1(0) = a + b + 3c$$
$$2 = y_2(0) = b + c$$
$$1 = y_3(0) = a + c.$$

Solving this system gives $a = 4$, $b = 5$, and $c = -3$. Thus the solution to the given system of differential equations and initial conditions is

$$y_1 = 4e^{-t} + 5e^t - 9e^{2t}$$
$$y_2 = \phantom{4e^{-t} +} 5e^t - 3e^{2t}$$
$$y_3 = 4e^{-t} - 3e^{2t}.$$

39. Let $y_1 = y$, $y_2 = y'$, and $y_3 = y''$. Then the given equation can be written

$$y''' = -2y + y' + 2y''$$

or

$$y_3' = -2y_1 + y_2 + 2y_3.$$

So the given equation is equivalent to the system

$$y_1' = y_2$$
$$y_2' = y_3$$
$$y_3' = -2y_1 + y_2 + 2y_3.$$

Write this system in the matrix form $\mathbf{y}' = A\mathbf{y}$, where

$$A = \begin{bmatrix} 0 & 1 & 0 \\ 0 & 0 & 1 \\ -2 & 1 & 2 \end{bmatrix}.$$

The characteristic polynomial of A is

$$\det (A - tI_3) = -t^3 + 2t^2 + t - 2 = -(t + 1)(t - 1)(t - 2),$$

129

and so the eigenvalues of A are -1, 1, and 2. Bases for the eigenspaces of A corresponding to the eigenvalues -1, 1, and 2 are

$$\left\{ \begin{bmatrix} 1 \\ -1 \\ 1 \end{bmatrix} \right\}, \quad \left\{ \begin{bmatrix} 1 \\ 1 \\ 1 \end{bmatrix} \right\}, \quad \text{and} \left\{ \begin{bmatrix} 1 \\ 2 \\ 4 \end{bmatrix} \right\},$$

respectively. Hence $A = PDP^{-1}$, where

$$P = \begin{bmatrix} 1 & 1 & 1 \\ -1 & 1 & 2 \\ 1 & 1 & 4 \end{bmatrix} \quad \text{and} \quad D = \begin{bmatrix} -1 & 0 & 0 \\ 0 & 1 & 0 \\ 0 & 0 & 2 \end{bmatrix}.$$

The solution of $\mathbf{z}' = D\mathbf{z}$ is

$$\begin{aligned} z_1 &= ae^{-t} \\ z_2 &= be^{t} \\ z_3 &= ce^{2t}. \end{aligned}$$

Thus the general solution to the original system is

$$\begin{bmatrix} y_1 \\ y_2 \\ y_3 \end{bmatrix} = \mathbf{y} = P\mathbf{z} = \begin{bmatrix} 1 & 1 & 1 \\ -1 & 1 & 2 \\ 1 & 1 & 4 \end{bmatrix} \begin{bmatrix} ae^{-t} \\ be^{t} \\ ce^{2t} \end{bmatrix}$$

$$= \begin{bmatrix} ae^{-t} + be^{t} + ce^{2t} \\ -ae^{-t} + be^{t} + 2ce^{2t} \\ ae^{-t} + be^{t} + 4ce^{2t} \end{bmatrix}.$$

Taking $t = 0$, we have

$$\begin{aligned} 2 &= y(0) = y_1(0) = a + b + c \\ -3 &= y'(0) = y_2(0) = -a + b + 2c \\ 5 &= y''(0) = y_3(0) = a + b + 4c. \end{aligned}$$

Solving this system yields $a = 3$, $b = -2$, and $c = 1$. Hence the particular solution to the original system of differential equations and initial conditions is

$$\begin{aligned} y_1 &= 3e^{-t} - 2e^{t} + e^{2t} \\ y_2 &= -3e^{-t} - 2e^{t} + 2e^{2t} \\ y_3 &= 3e^{-t} - 2e^{t} + 4e^{2t}. \end{aligned}$$

Therefore the particular solution to the given third-order differential equation and initial conditions is

$$y(t) = y_1(t) = 3e^{-t} - 2e^{t} + e^{2t}.$$

41. Take $w = 10$ lbs, $b = 0.625$, and $k = 1.25$ lbs/foot in the equation

$$\frac{w}{g} y''(t) + by'(t) + ky(t) = 0.$$

Then the equation simplifies to the form

$$y'' + 2y' + 4y = 0.$$

We transform this differential equation into a system of differential equations by letting $y_1 = y$ and $y_2 = y'$. These substitutions produce the system

$$\begin{aligned} y_1' &= \quad\quad\; y_2 \\ y_2' &= -4y_1 - 2y_2, \end{aligned}$$

which can be written in the matrix form $\mathbf{y}' = A\mathbf{y}$, where

$$A = \begin{bmatrix} 0 & 1 \\ -4 & -2 \end{bmatrix}.$$

The characteristic polynomial of A is $t^2 + 2t + 4$, which has the nonreal roots $-1 + \sqrt{3}\,i$ and $-1 - \sqrt{3}\,i$. The general solution to the original differential equation can be written as

$$\begin{aligned} y &= ae^{(-1-\sqrt{3}\,i)t} + be^{(-1+\sqrt{3}\,i)t} \\ &= ae^{-t}(\cos\sqrt{3}\,t + i\sin\sqrt{3}\,t) + be^{-t}(\cos\sqrt{3}\,t - i\sin\sqrt{3}\,t) \end{aligned}$$

using Euler's formula, or, equivalently, as

$$y = ce^{-t}\cos\sqrt{3}\,t + de^{-t}\sin\sqrt{3}\,t.$$

45. The differential equation $y''' + ay'' + by' + cy = 0$ can be written as $\mathbf{y}' = A\mathbf{y}$, where

$$A = \begin{bmatrix} 0 & 1 & 0 \\ 0 & 0 & 1 \\ -c & -b & -a \end{bmatrix}.$$

By Exercise 44, the characteristic polynomial of A is $-t^3 - at^2 - bt - c$, and so $\lambda_i^3 = -a\lambda_i^2 - b\lambda_i - c$ for $i = 1, 2, 3$. Now

$$A\begin{bmatrix} 1 \\ \lambda_i \\ \lambda_i^2 \end{bmatrix} = \begin{bmatrix} \lambda_i \\ \lambda_i^2 \\ -c - b\lambda_i - a\lambda_i^2 \end{bmatrix} = \begin{bmatrix} \lambda_i \\ \lambda_i^2 \\ \lambda_i^3 \end{bmatrix} = \lambda_i \begin{bmatrix} 1 \\ \lambda_i \\ \lambda_i^2 \end{bmatrix}.$$

Thus

$$\mathbf{v}_i = \begin{bmatrix} 1 \\ \lambda_i \\ \lambda_i^2 \end{bmatrix}$$

is an eigenvector of A with λ_i as its corresponding eigenvalue. So $\{v_1, v_2, v_3\}$ is a basis for $\mathcal{R}^3$ consisting of eigenvectors of A. Thus we may use the boxed solution to $\mathbf{y}' = A\mathbf{y}$ on page 296 with

$$P = [\mathbf{v}_1\ \mathbf{v}_2\ \mathbf{v}_3] \qquad \text{and} \qquad D = \begin{bmatrix} \lambda_1 & 0 & 0 \\ 0 & \lambda_2 & 0 \\ 0 & 0 & \lambda_3 \end{bmatrix}.$$

49. The given difference equation can be written as $\mathbf{s}_n = A\mathbf{s}_{n-1}$, where

$$\mathbf{s}_n = \begin{bmatrix} r_{n+1} \\ r_n \end{bmatrix} \qquad \text{and} \qquad A = \begin{bmatrix} 3 & 4 \\ 1 & 0 \end{bmatrix}.$$

Taking

$$P = \begin{bmatrix} -1 & 4 \\ 1 & 1 \end{bmatrix} \qquad \text{and} \qquad D = \begin{bmatrix} -1 & 0 \\ 0 & 4 \end{bmatrix},$$

we have $A = PDP^{-1}$. Hence $A^n = PD^nP^{-1}$, and so

$$\mathbf{s}_n = A^n\mathbf{s}_0 = PD^nP^{-1}\mathbf{s}_0$$

$$= \begin{bmatrix} -1 & 4 \\ 1 & 1 \end{bmatrix} \begin{bmatrix} (-1)^n & 0 \\ 0 & 4^n \end{bmatrix} \begin{bmatrix} -.2 & .8 \\ .2 & .2 \end{bmatrix} \begin{bmatrix} 1 \\ 1 \end{bmatrix}$$

$$= \begin{bmatrix} (-1)^{n+1} & 0 \\ (-1)^n & 4^n \end{bmatrix} \begin{bmatrix} .6 \\ .4 \end{bmatrix}$$

$$= \begin{bmatrix} .6(-1)^{n+1} \\ .6(-1)^n + .4(4^n) \end{bmatrix}.$$

Equating the second components of $\mathbf{s}_n$ and the previous vector, we have

$$r_n = .6(-1)^n + .4(4^n) \text{ for } n \geq 0.$$

Hence

$$r_6 = .6(-1)^6 + .4(4^6) = 1639.$$

53. Since the given difference equation is of the third order, $\mathbf{s}_n$ is a vector in $\mathcal{R}^3$ and A is a 3×3 matrix. Taking

$$\mathbf{s}_n = \begin{bmatrix} r_{n+2} \\ r_{n+1} \\ r_n \end{bmatrix} \qquad \text{and} \qquad A = \begin{bmatrix} 4 & -2 & 5 \\ 1 & 0 & 0 \\ 0 & 1 & 0 \end{bmatrix},$$

we see that the matrix form of the given equation is $\mathbf{s}_n = A\mathbf{s}_{n-1}$.

57. We have

$$A\mathbf{s}_0 = \begin{bmatrix} 1 & 0 \\ c & a \end{bmatrix} \begin{bmatrix} 1 \\ r_0 \end{bmatrix} = \begin{bmatrix} 1 \\ c + ar_0 \end{bmatrix} = \begin{bmatrix} 1 \\ r_1 \end{bmatrix} = \mathbf{s}_1,$$

$$A\mathbf{s}_1 = \begin{bmatrix} 1 & 0 \\ c & a \end{bmatrix} \begin{bmatrix} 1 \\ r_1 \end{bmatrix} = \begin{bmatrix} 1 \\ c + ar_1 \end{bmatrix} = \begin{bmatrix} 1 \\ r_2 \end{bmatrix} = \mathbf{s}_2,$$

and, in general,

$$A\mathbf{s}_{n-1} = \begin{bmatrix} 1 & 0 \\ c & a \end{bmatrix} \begin{bmatrix} 1 \\ r_{n-1} \end{bmatrix} = \begin{bmatrix} 1 \\ c + ar_{n-1} \end{bmatrix} = \begin{bmatrix} 1 \\ r_n \end{bmatrix} = \mathbf{s}_n.$$

Hence

$$\mathbf{s}_n = A\mathbf{s}_{n-1} = A(A\mathbf{s}_{n-2}) = A^2\mathbf{s}_{n-2} = A^3\mathbf{s}_{n-3} = \cdots = A^n\mathbf{s}_0.$$

61. The given system of differential equations has the form $\mathbf{y}' = A\mathbf{y}$, where

$$\mathbf{y} = \begin{bmatrix} y_1 \\ y_2 \\ y_3 \\ y_4 \end{bmatrix} \quad \text{and} \quad A = \begin{bmatrix} 3.2 & 4.1 & 7.7 & 3.7 \\ -0.3 & 1.2 & 0.2 & 0.5 \\ -1.8 & -1.8 & -4.4 & -1.8 \\ 1.7 & -0.7 & 2.9 & 0.4 \end{bmatrix}.$$

The characteristic polynomial of A is

$$t^4 - 0.4t^3 - 0.79t^2 + .166t + 0.24 = (t + 0.8)(t + 0.1)(t - 0.3)(t - 1).$$

Since A has four distinct eigenvalues, it is diagonalizable; in fact, $A = PDP^{-1}$, where

$$P = \begin{bmatrix} 1 & -1 & -1 & 2 \\ 0 & -1 & -2 & -1 \\ -1 & 0 & 0 & -1 \\ 1 & 2 & 3 & 2 \end{bmatrix} \quad \text{and} \quad D = \begin{bmatrix} -0.8 & 0.0 & 0.0 & 0 \\ 0.0 & -0.1 & 0.0 & 0 \\ 0.0 & 0.0 & 0.3 & 0 \\ 0.0 & 0.0 & 0.0 & 1 \end{bmatrix}.$$

The solution to $\mathbf{z}' = D\mathbf{z}$ is

$$\begin{bmatrix} z_1 \\ z_2 \\ z_3 \\ z_4 \end{bmatrix} = \begin{bmatrix} ae^{-0.8t} \\ be^{-0.1t} \\ ce^{0.3t} \\ de^t \end{bmatrix}.$$

Hence the general solution to the original equation is

$$\begin{bmatrix} y_1 \\ y_2 \\ y_3 \\ y_4 \end{bmatrix} = \mathbf{y} = P\mathbf{z} = \begin{bmatrix} ae^{-0.8t} - be^{-0.1t} - ce^{0.3t} + 2de^t \\ -be^{-0.1t} - 2ce^{0.3t} - de^t \\ -ae^{-0.8t} \qquad\qquad\qquad - de^t \\ ae^{-0.8t} + 2be^{-0.1t} + 3ce^{0.3t} + 2de^t \end{bmatrix}.$$

When $t = 0$, the preceding equation takes the form

$$\begin{bmatrix} 1 \\ -4 \\ 2 \\ 3 \end{bmatrix} = P \begin{bmatrix} a \\ b \\ c \\ d \end{bmatrix},$$

and so $a = -6$, $b = 2$, $c = -1$, and $d = 4$. Thus the particular solution of the original system that satisfies the given initial condition is

$$\begin{aligned} y_1 &= -6e^{-0.8t} - 2e^{-0.1t} + e^{0.3t} + 8e^{t} \\ y_2 &= \qquad\qquad -2e^{-0.1t} + 2e^{0.3t} - 4e^{t} \\ y_3 &= \quad 6e^{-0.8t} \qquad\qquad\qquad - 4e^{t} \\ y_4 &= -6e^{-0.8t} + 4e^{-0.1t} - 3e^{0.3t} + 8e^{t}. \end{aligned}$$

CHAPTER 5 REVIEW

1. (a) True
   (b) False, there are infinitely many eigenvectors that correspond to a particular eigenvalue.
   (c) True
   (d) True
   (e) True
   (f) True
   (g) False, the linear operator on $\mathcal{R}^2$ that rotates a vector by $90°$ has no real eigenvalues.
   (h) False, the rotation matrix $A_{90°}$ has no (real) eigenvalues.
   (i) False, I_n has only one eigenvalue, namely, 1.
   (j) False, if two $n \times n$ matrices have the same characteristic polynomial, they have the same *eigenvalues*.
   (k) True
   (l) True
   (m) False, if $A = PDP^{-1}$, where P is an invertible matrix and D is a diagonal matrix, then the columns of P are a basis for $\mathcal{R}^n$ consisting of eigenvectors of A.
   (n) True
   (o) True
   (p) True

5. The characteristic polynomial of the given matrix A is

$$-t^3 - 5t^2 - 8t - 4 = -(t + 1)(t + 2)^2.$$

So the eigenvalues of A are -1 and -2 (with multiplicity 2). A basis for the eigenspace of A corresponding to eigenvalue -1 is obtained from the parametric representation

134

of the general solution to $(A - I_3)\mathbf{x} = \mathbf{0}$. Such a basis is

$$\left\{ \begin{bmatrix} 0 \\ 1 \\ 0 \end{bmatrix} \right\}.$$

In a similar manner, a basis for the eigenspace of A corresponding to eigenvalue -2 is obtained from the parametric representation of the general solution to $(A + 2I_3)\mathbf{x} = \mathbf{0}$. Such a basis is

$$\left\{ \begin{bmatrix} -1 \\ 1 \\ 0 \end{bmatrix} \right\}.$$

9. The characteristic polynomial of the given matrix A is $-t^3 - t^2 + t + 1 = -(t-1)(t+1)^2$, and so the eigenvalues of A are 1 and -1 (with multiplicity 2). Since the reduced row echelon form of $A + I_3$ is

$$\begin{bmatrix} 1 & 0 & 0 \\ 0 & 1 & 1 \\ 0 & 0 & 0 \end{bmatrix},$$

the eigenspace of A corresponding to eigenvalue -1 has dimension $3 - \operatorname{rank}(A + 3I_3) = 3 - 2 = 1$. Since this eigenvalue has multiplicity 2, A is not diagonalizable.

13. The standard matrix of T is

$$A = \begin{bmatrix} 2 & 0 & 0 \\ 0 & 2 & 0 \\ -3 & 3 & -1 \end{bmatrix}.$$

The characteristic polynomial of A is $-t^3 + 3t^2 - 4 = -(t+1)(t-2)^2$, and so A has eigenvalues of -1 and 2 (with multiplicity 2). Bases for the corresponding eigenspaces of A (and T) are

$$\mathcal{B}_1 = \left\{ \begin{bmatrix} 0 \\ 0 \\ 1 \end{bmatrix} \right\} \qquad \text{and} \qquad \mathcal{B}_2 = \left\{ \begin{bmatrix} 1 \\ 1 \\ 0 \end{bmatrix}, \begin{bmatrix} -1 \\ 0 \\ 1 \end{bmatrix} \right\}.$$

Then $\mathcal{B}_1 \cup \mathcal{B}_2$ is a basis for $\mathcal{R}^3$ consisting of eigenvectors of T.

17. The eigenvalues of the given matrix are c, 2, and -2. Since an $n \times n$ matrix having n distinct eigenvalues is diagonalizable, A is diagonalizable if $c \neq 2$ and $c \neq -2$. For $c = 2$, the reduced row echelon form of $A - 2I_3$ is

$$\begin{bmatrix} 0 & 1 & 0 \\ 0 & 0 & 1 \\ 0 & 0 & 0 \end{bmatrix}.$$

135

So the eigenspace corresponding to eigenvalue 2 has dimension 1. Hence A is not diagonalizable if $c = 2$. Likewise, for $c = -2$, the reduced row echelon form of $A + 2I_3$ is

$$\begin{bmatrix} 0 & 1 & 0 \\ 0 & 0 & 1 \\ 0 & 0 & 0 \end{bmatrix},$$

and so A is not diagonalizable for $c = -2$.

19. The characteristic polynomial of A is

$$\det(A - tI_2) = t^2 - t - 2 = (t+1)(t-2);$$

so the eigenvalues of A are -1 and 2. Bases for the eigenspaces of A corresponding to the eigenvalues -1 and 2 are

$$\left\{ \begin{bmatrix} 1 \\ 1 \end{bmatrix} \right\} \quad \text{and} \quad \left\{ \begin{bmatrix} 2 \\ 1 \end{bmatrix} \right\}.$$

Hence $A = PDP^{-1}$, where

$$P = \begin{bmatrix} 1 & 2 \\ 1 & 1 \end{bmatrix} \quad \text{and} \quad D = \begin{bmatrix} -1 & 0 \\ 0 & 2 \end{bmatrix}.$$

So, for any positive integer k,

$$A^k = PD^kP^{-1} = \begin{bmatrix} 1 & 2 \\ 1 & 1 \end{bmatrix} \begin{bmatrix} (-1)^k & 0 \\ 0 & 2^k \end{bmatrix} \begin{bmatrix} 1 & 2 \\ 1 & 1 \end{bmatrix}^{-1}$$

$$= \begin{bmatrix} 1 & 2 \\ 1 & 1 \end{bmatrix} \begin{bmatrix} -(-1)^k & 2(-1)^k \\ 2^k & -2^k \end{bmatrix}$$

$$= \begin{bmatrix} 2^{k+1} - (-1)^k & 2(-1)^k - 2^{k+1} \\ 2^k - (-1)^k & 2(-1)^k - 2^k \end{bmatrix}.$$

21. A basis $\mathcal{B}$ such that $[T]_{\mathcal{B}}$ is a diagonal matrix is a basis for $\mathcal{R}^3$ consisting of eigenvectors of T; so we proceed as in Exercise 13. The standard matrix of T is

$$A = \begin{bmatrix} -4 & -3 & -3 \\ 0 & -1 & 0 \\ 6 & 6 & 5 \end{bmatrix},$$

and so its characteristic polynomial is $-t^3 + 3t + 2 = -(t-2)(t+1)^2$. Thus the eigenvalues of T are 2 and -1 (with multiplicity 2). Solving $(A - 2I_3)\mathbf{x} = \mathbf{0}$, we obtain the following basis for the eigenspace of T corresponding to eigenvalue 2:

$$\mathcal{B}_1 = \left\{ \begin{bmatrix} -1 \\ 0 \\ 2 \end{bmatrix} \right\}.$$

Solving $(A + I_3)\mathbf{x} = \mathbf{0}$, we obtain the following basis for the eigenspace of T corresponding to eigenvalue -1:

$$\mathcal{B}_2 = \left\{ \begin{bmatrix} -1 \\ 1 \\ 0 \end{bmatrix}, \begin{bmatrix} -1 \\ 0 \\ 1 \end{bmatrix} \right\}.$$

Then $\mathcal{B}_1 \cup \mathcal{B}_2$ is a basis for $\mathcal{R}^3$ consisting of eigenvectors of T, and hence $[T]_\mathcal{B}$ is a diagonal matrix. In fact,

$$[T]_\mathcal{B} = \begin{bmatrix} 2 & 0 & 0 \\ 0 & -1 & 0 \\ 0 & 0 & -1 \end{bmatrix}.$$

25. If $I_n - A$ is invertible, then $\det (I_n - A) \neq 0$. Hence

$$\det (A - I_n) = (-1)^n \det (I_n - A) \neq 0,$$

and so 1 is not a root of $\det (A - tI_n)$. Conversely, if $I_n - A$ is not invertible, then

$$\det (A - I_n) = (-1)^n \det (I_n - A) = 0.$$

So 1 is a root of $\det (A - tI_n)$.

Chapter 6

Orthogonality

6.1 THE GEOMETRY OF VECTORS

1. (a) True
   (b) False, the dot product of two vectors is a scalar.
   (c) False, the norm is the square root of the dot product of a vector with itself.
   (d) False, the norm is the product of the *absolute value* of the multiple with the norm of the vector.
   (e) False, for example, if $\mathbf{v}$ is a nonzero vector, then

   $$\|\mathbf{v} + (-\mathbf{v})\| = 0 \neq \|\mathbf{v}\| + \| - \mathbf{v}\|.$$

   (f) True
   (g) True

5. $\|\mathbf{u}\| = \sqrt{1^2 + (-1)^2 + 3^2} = \sqrt{11}$, $\|\mathbf{v}\| = \sqrt{2^2 + 1^2 + 0^2} = \sqrt{5}$,
   and $d = \|\mathbf{u} - \mathbf{v}\| = \sqrt{(1-2)^2 + (-1-1)^2 + (3-0)^2} = \sqrt{14}$

9. $\mathbf{u} \cdot \mathbf{v} = (1)(2) + (-1)(1) = 1$, and hence $\mathbf{u}$ and $\mathbf{v}$ are not orthogonal because $\mathbf{u} \cdot \mathbf{v} \neq 0$.

13. -2, no

17. $\|\mathbf{u}\|^2 = 1^2 + 2^2 + 3^2 = 14$, $\|\mathbf{v}\|^2 = (-11)^2 + 4^2 + 1^2 = 138$,
   and $\|\mathbf{u} + \mathbf{v}\|^2 = (1 - 11)^2 + (2 + 4)^2 + (3 + 1)^2 = 152$. So

   $$\|\mathbf{u}\|^2 + \|\mathbf{v}\|^2 = 14 + 138 = 152 = \|\mathbf{u} + \mathbf{v}\|^2.$$

21. $\|\mathbf{u}\| = \sqrt{2^2 + (-1)^2 + 3^2} = \sqrt{14}$, $\|\mathbf{v}\| = \sqrt{4^2 + 0^2 + 1^2} = \sqrt{17}$,
   and $\|\mathbf{u} + \mathbf{v}\| = \sqrt{(2 + 4)^2 + (-1 + 0)^2 + (3 + 1)^2} = \sqrt{53}$. So

   $$\|\mathbf{u} + \mathbf{v}\| = \sqrt{53} \leq \sqrt{14} + \sqrt{17} = \|\mathbf{u}\| + \|\mathbf{v}\|.$$

25. $\|\mathbf{u}\| = \sqrt{4^2 + 2^2 + 1^2} = \sqrt{21}$, $\|\mathbf{v}\| = \sqrt{2^2 + (-1)^2 + (-1)^2} = \sqrt{6}$,
   and $\mathbf{u} \cdot \mathbf{v} = (4)(2) + 2(-1) + (1)(-1) - 5$. So

   $$|\mathbf{u} \cdot \mathbf{v}| = 5 \leq \sqrt{21}\sqrt{6} = \|\mathbf{u}\|\|\mathbf{v}\|.$$

29. Let $\mathbf{u} = \begin{bmatrix} 1 \\ 3 \end{bmatrix}$, a nonzero vector that lies along the line $y = 3x$. Then $\mathbf{w} = c\mathbf{u}$, where

$$c = \frac{\mathbf{u} \cdot \mathbf{v}}{\mathbf{u} \cdot \mathbf{u}} = \frac{7}{10},$$

and hence $\mathbf{w} = \begin{bmatrix} 0.7 \\ 2.1 \end{bmatrix}$. Therefore

$$d = \|\mathbf{v} - \mathbf{w}\| = \sqrt{(4 - 0.7)^2 + (1 - 2.1)^2} = 1.1\sqrt{10}.$$

33. $\|\mathbf{u} + \mathbf{v}\|^2 = (\mathbf{u} + \mathbf{v}) \cdot (\mathbf{u} + \mathbf{v}) = \mathbf{u} \cdot \mathbf{u} + 2\mathbf{u} \cdot \mathbf{v} + \mathbf{v} \cdot \mathbf{v} = 4 - 2 + 9 = 11$

37. Suppose that $\mathbf{u} \cdot \mathbf{u} = 0$, where $\mathbf{u}$ is in $\mathcal{R}^n$. Then $u_1^2 + u_2^2 + \cdots + u_n^2 = 0$. Since $u_i^2 \geq 0$ for all i, we have that $u_i^2 = 0$ for all i. It follows that $u_i = 0$ for each i, and hence $\mathbf{u} = \mathbf{0}$.

Conversely, suppose that $\mathbf{u} = \mathbf{0}$. Then $\mathbf{u} \cdot \mathbf{u} = 0^2 + 0^2 + \cdots + 0^2 = 0$.

41. We have

$$\mathbf{w} \cdot \mathbf{z} = \mathbf{w} \cdot (\mathbf{v} - \mathbf{w})$$
$$= \mathbf{w} \cdot \mathbf{v} - \mathbf{w} \cdot \mathbf{w}$$
$$= \frac{(\mathbf{u} \cdot \mathbf{v})^2}{\|\mathbf{u}\|^2} - \frac{(\mathbf{u} \cdot \mathbf{v})^2}{\|\mathbf{u}\|^4} \mathbf{u} \cdot \mathbf{u}$$
$$= \frac{(\mathbf{u} \cdot \mathbf{v})^2}{\|\mathbf{u}\|^2} - \frac{(\mathbf{u} \cdot \mathbf{v})^2}{\|\mathbf{u}\|^2}$$
$$= 0.$$

45. $(\mathbf{u} + \mathbf{v}) \cdot \mathbf{w} = \mathbf{w} \cdot (\mathbf{u} + \mathbf{v}) = \mathbf{w} \cdot \mathbf{u} + \mathbf{w} \cdot \mathbf{v} = \mathbf{u} \cdot \mathbf{w} + \mathbf{v} \cdot \mathbf{w}$

49. We have

$$\|\mathbf{u} + \mathbf{v}\|^2 + \|\mathbf{u} - \mathbf{v}\|^2 = \left(\|\mathbf{u}\|^2 + 2\mathbf{u} \cdot \mathbf{v} + \|\mathbf{v}\|^2\right) + \left(\|\mathbf{u}\|^2 - 2\mathbf{u} \cdot \mathbf{v} + \|\mathbf{v}\|^2\right)$$
$$= 2\|\mathbf{u}\|^2 + 2\|\mathbf{v}\|^2.$$

53. (d) $\|\mathbf{u} + \mathbf{v}\| = \|\mathbf{u}\| + \|\mathbf{v}\|$ if and only if $\mathbf{u}$ is a nonnegative multiple of $\mathbf{v}$ or $\mathbf{v}$ is a nonnegative multiple of $\mathbf{u}$.

(e) If two sides of a triangle are parallel, then the triangle is "degenerate", that is, it contains no region of positive area, and the third side coincides with the union of the other two sides.

6.2   ORTHOGONAL VECTORS

1. (a)   False, if $\mathbf{0}$ lies in the set, then the set is linearly dependent.
   (b)   True
   (c)   False if S is not a subspace of $\mathcal{R}^n$.
   (d)   False, if F is a finite set and G is the span of F, then $F^{\perp} = G^{\perp}$, but $F \neq G$.
   (e)   True
   (f)   True

5. We have

$$\begin{bmatrix} 1 \\ 2 \\ 3 \\ -1 \end{bmatrix} \cdot \begin{bmatrix} 1 \\ 1 \\ -1 \\ 0 \end{bmatrix} = (1)(1) + (2)(1) + (3)(-1) + (-1)(0) = 0,$$

$$\begin{bmatrix} 1 \\ 2 \\ 3 \\ -1 \end{bmatrix} \cdot \begin{bmatrix} 3 \\ -3 \\ 0 \\ -1 \end{bmatrix} = (1)(3) + (2)(-3) + (3)(0) + (-1)(-1) = 0,$$

and

$$\begin{bmatrix} 1 \\ 1 \\ -1 \\ 0 \end{bmatrix} \cdot \begin{bmatrix} 3 \\ -3 \\ 0 \\ -1 \end{bmatrix} = (1)(3) + (1)(-3) + (-1)(0) + (0)(-1) = 0.$$

Therefore the set is orthogonal.

9. (a) Let

$$\mathbf{u}_1 = \begin{bmatrix} 0 \\ 1 \\ 1 \\ 1 \end{bmatrix}, \qquad \mathbf{u}_2 = \begin{bmatrix} 1 \\ 0 \\ 1 \\ 1 \end{bmatrix}, \qquad \text{and} \qquad \mathbf{u}_3 = \begin{bmatrix} 1 \\ 1 \\ 0 \\ 1 \end{bmatrix}.$$

Set

$$\mathbf{v}_1 = \mathbf{u}_1,$$

$$\mathbf{v}_2 = \mathbf{u}_2 - \frac{\mathbf{u}_2 \cdot \mathbf{v}_1}{\|\mathbf{v}_1\|^2} \mathbf{v}_1$$

$$= \begin{bmatrix} 1 \\ 0 \\ 1 \\ 1 \end{bmatrix} - \frac{2}{3} \begin{bmatrix} 0 \\ 1 \\ 1 \\ 1 \end{bmatrix} = \frac{1}{3} \begin{bmatrix} 3 \\ -2 \\ 1 \\ 1 \end{bmatrix},$$

and

$$\mathbf{v}_3 = \mathbf{u}_3 - \frac{\mathbf{u}_3 \cdot \mathbf{v}_1}{\|\mathbf{v}_1\|^2}\mathbf{v}_1 - \frac{\mathbf{u}_3 \cdot \mathbf{v}_2}{\|\mathbf{v}_2\|^2}\mathbf{v}_2$$

$$= \begin{bmatrix} 1 \\ 1 \\ 0 \\ 1 \end{bmatrix} - \frac{2}{3}\begin{bmatrix} 0 \\ 1 \\ 1 \\ 1 \end{bmatrix} - \left(\frac{\frac{2}{3}}{\frac{5}{3}}\right)\left(\frac{1}{3}\right)\begin{bmatrix} 3 \\ -2 \\ 1 \\ 1 \end{bmatrix}$$

$$= \frac{1}{5}\begin{bmatrix} 3 \\ 3 \\ -4 \\ 1 \end{bmatrix}.$$

Thus

$$\{\mathbf{v}_1, \mathbf{v}_2, \mathbf{v}_2\} = \left\{ \begin{bmatrix} 0 \\ 1 \\ 1 \\ 1 \end{bmatrix}, \frac{1}{3}\begin{bmatrix} 3 \\ -2 \\ 1 \\ 1 \end{bmatrix}, \frac{1}{5}\begin{bmatrix} 3 \\ 3 \\ -4 \\ 1 \end{bmatrix} \right\}$$

is the corresponding orthogonal set.

(b) Using the notation in (a), we obtain the orthonormal set

$$\left\{ \frac{1}{\|\mathbf{v}_1\|}\mathbf{v}_1, \frac{1}{\|\mathbf{v}_2\|}\mathbf{v}_2, \frac{1}{\|\mathbf{v}_1\|}\mathbf{v}_3 \right\} = \left\{ \frac{1}{\sqrt{3}}\begin{bmatrix} 0 \\ 1 \\ 1 \\ 1 \end{bmatrix}, \frac{1}{\sqrt{15}}\begin{bmatrix} 3 \\ -2 \\ 1 \\ 1 \end{bmatrix}, \frac{1}{\sqrt{35}}\begin{bmatrix} 3 \\ 3 \\ -4 \\ 1 \end{bmatrix} \right\}.$$

13. We have

$$\mathbf{v} = \frac{\begin{bmatrix} 1 \\ 8 \end{bmatrix} \cdot \begin{bmatrix} 2 \\ 1 \end{bmatrix}}{\left\|\begin{bmatrix} 2 \\ 1 \end{bmatrix}\right\|^2}\begin{bmatrix} 2 \\ 1 \end{bmatrix} + \frac{\begin{bmatrix} 1 \\ 8 \end{bmatrix} \cdot \begin{bmatrix} -1 \\ 2 \end{bmatrix}}{\left\|\begin{bmatrix} -1 \\ 2 \end{bmatrix}\right\|^2}\begin{bmatrix} -1 \\ 2 \end{bmatrix}$$

$$= \frac{10}{5}\begin{bmatrix} 2 \\ 1 \end{bmatrix} + \frac{15}{5}\begin{bmatrix} -1 \\ 2 \end{bmatrix}$$

$$= 2\begin{bmatrix} 2 \\ 1 \end{bmatrix} + 3\begin{bmatrix} -1 \\ 2 \end{bmatrix}.$$

17. A vector $\mathbf{v}$ is in $\mathcal{S}^\perp$ if and only if

$$\mathbf{v} \cdot \begin{bmatrix} 1 \\ -1 \\ 2 \end{bmatrix} = v_1 - v_2 + 2v_3 = 0.$$

The solution set to this system has a basis $\left\{ \begin{bmatrix} 1 \\ 1 \\ 0 \end{bmatrix}, \begin{bmatrix} -2 \\ 0 \\ 1 \end{bmatrix} \right\}$.

21. (a) We have

$$\mathbf{w} = \begin{bmatrix} 1 \\ 3 \end{bmatrix} \cdot \left(\frac{1}{\sqrt{2}} \begin{bmatrix} 1 \\ -1 \end{bmatrix} \right) \left(\frac{1}{\sqrt{2}} \begin{bmatrix} 1 \\ -1 \end{bmatrix} \right) = \frac{-2}{2} \begin{bmatrix} 1 \\ -1 \end{bmatrix} = \begin{bmatrix} -1 \\ 1 \end{bmatrix}$$

and

$$\mathbf{z} = \mathbf{v} - \mathbf{w} = \begin{bmatrix} 1 \\ 3 \end{bmatrix} - \begin{bmatrix} -1 \\ 1 \end{bmatrix} = \begin{bmatrix} 2 \\ 2 \end{bmatrix}.$$

(b) $\mathbf{w} = \begin{bmatrix} -1 \\ 1 \end{bmatrix}$

(c) $\|\mathbf{z}\| = \left\| \begin{bmatrix} 2 \\ 2 \end{bmatrix} \right\| = \sqrt{8}$

25. We divide the vector in the given basis by the norm of the vector, 5, to obtain an orthonormal basis

$$\left\{ \frac{1}{5} \begin{bmatrix} -3 \\ 4 \end{bmatrix} \right\}$$

for W. Thus

$$\mathbf{w} = \left(\begin{bmatrix} -10 \\ 5 \end{bmatrix} \cdot \left(\frac{1}{5} \begin{bmatrix} -3 \\ 4 \end{bmatrix} \right) \right) \left(\frac{1}{5} \begin{bmatrix} -3 \\ 4 \end{bmatrix} \right) = \frac{50}{25} \begin{bmatrix} -3 \\ 4 \end{bmatrix} = \begin{bmatrix} -6 \\ 8 \end{bmatrix},$$

and

$$\mathbf{z} = \mathbf{v} - \mathbf{w} = \begin{bmatrix} -10 \\ 5 \end{bmatrix} - \begin{bmatrix} -6 \\ 8 \end{bmatrix} = \begin{bmatrix} -4 \\ -3 \end{bmatrix}.$$

The orthogonal projection of $\mathbf{v}$ on W is $\mathbf{w} = \begin{bmatrix} -6 \\ 8 \end{bmatrix}$, and the distance from $\mathbf{v}$ to $\mathbf{w}$ is

$\|\mathbf{z}\| = \left\| \begin{bmatrix} -4 \\ -3 \end{bmatrix} \right\| = 5.$

29. For any $i \neq j$,

$$(c_i \mathbf{v}_i) \cdot (c_j \mathbf{v}_j) = (c_i c_j)(\mathbf{v}_i \cdot \mathbf{v}_j) = (c_i c_j) \cdot 0 = 0,$$

and hence $c_i \mathbf{v}_i$ and $c_j \mathbf{v}_j$ are orthogonal.

33. First, note that $\dim W = k$, and hence by Exercise 32,

$$\dim W^{\perp} = n - \dim W = n - k.$$

Next observe that each $\mathbf{v}_i$ for $i > k$ is orthogonal to $\{\mathbf{v}_1, \mathbf{v}_2, \ldots, \mathbf{v}_k\}$, which is a spanning set for W, and hence lies in $W^{\perp}$ by Exercise 31. Since $\{\mathbf{v}_{k+1}, \mathbf{v}_{k+2}, \ldots, \mathbf{v}_n\}$ is orthogonal and consists of nonzero vectors, it is a linearly independent subset of $W^{\perp}$. Also, it contains $n - k$ vectors. Therefore this set is a basis for $W^{\perp}$.

37. Suppose that $\mathbf{v}$ is in $W^{\perp}$. Then $\mathbf{v}$ is orthogonal to every vector in W. Since V is a subset of W, $\mathbf{v}$ is orthogonal to every vector in V. Therefore $\mathbf{v}$ is in $V^{\perp}$, and we conclude that $W^{\perp}$ is contained in $V^{\perp}$.

41. By Theorem 4.4, $\{\mathbf{v}_1, \mathbf{v}_2, \ldots, \mathbf{v}_k\}$ can be extended to a basis

$$\{\mathbf{v}_1, \mathbf{v}_2, \ldots, \mathbf{v}_k, \mathbf{u}_{k+1}, \ldots, \mathbf{u}_n\}$$

for $\mathcal{R}^n$. Applying the Gram-Schmidt process to this basis results in an orthogonal basis $\{\mathbf{v}_1, \mathbf{v}_2, \ldots, \mathbf{v}_k, \mathbf{w}_{k+1}, \ldots, \mathbf{w}_n\}$. (Note that by Exercise 30, the first k vectors of this new basis remain unchanged when applying the Gram-Schmidt process.) Finally, we replace each $\mathbf{w}_i$, $k + 1 \le i \le n$, by $\mathbf{v}_i = \frac{1}{\|\mathbf{w}_i\|} \mathbf{w}_i$ to obtain the desired orthonormal basis for $\mathcal{R}^n$.

6.3 LEAST-SQUARES APPROXIMATION AND ORTHOGONAL PROJECTION MATRICES

1. (a) True
   (b) False, the only $n \times n$ orthogonal projection matrix that is invertible is I_n.
   (c) True
   (d) False, it is the line that minimizes the sum of the *squares* of the vertical distances from the data points to the line.
   (e) True

5. Let

$$C = \begin{bmatrix} 1 & 0 \\ 1 & 1 \\ 0 & 0 \\ 1 & 1 \end{bmatrix},$$

the matrix whose columns are the vectors in the given basis for W. Then

$$P_W = C(C^T C)^{-1} C^T$$

$$= \begin{bmatrix} 1 & 0 \\ 1 & 1 \\ 0 & 0 \\ 1 & 1 \end{bmatrix} \begin{bmatrix} 3 & 2 \\ 2 & 2 \end{bmatrix}^{-1} \begin{bmatrix} 1 & 1 & 0 & 1 \\ 0 & 1 & 0 & 1 \end{bmatrix}$$

$$= \frac{1}{2} \begin{bmatrix} 1 & 0 \\ 1 & 1 \\ 0 & 0 \\ 1 & 1 \end{bmatrix} \begin{bmatrix} 2 & -2 \\ -2 & 3 \end{bmatrix} \begin{bmatrix} 1 & 1 & 0 & 1 \\ 0 & 1 & 0 & 1 \end{bmatrix}$$

$$= \frac{1}{2} \begin{bmatrix} 2 & 0 & 0 & 0 \\ 0 & 1 & 0 & 1 \\ 0 & 0 & 0 & 0 \\ 0 & 1 & 0 & 1 \end{bmatrix}.$$

9. The reduced row echelon form of the augmented matrix of this system is

$$\begin{bmatrix} 1 & 0 & -5 & 0 \\ 0 & 1 & 4 & 0 \end{bmatrix},$$

and hence the general solution is given by

$$\begin{array}{rl} x_1 = & 5x_3 \\ x_2 = & -4x_3 \\ x_3 & \text{free.} \end{array}$$

Thus the parametric representation of the general solution is

$$\begin{bmatrix} x_1 \\ x_2 \\ x_3 \end{bmatrix} = \begin{bmatrix} 5x_3 \\ -4x_3 \\ x_3 \end{bmatrix} = x_3 \begin{bmatrix} 5 \\ -4 \\ 1 \end{bmatrix},$$

and so

$$\left\{ \begin{bmatrix} 5 \\ -4 \\ 1 \end{bmatrix} \right\}$$

is a basis for the solution space. Let

$$C = \begin{bmatrix} 5 \\ -4 \\ 1 \end{bmatrix}.$$

Then

$$P_W = C(C^T C)^{-1} C^T$$

$$= \begin{bmatrix} 5 \\ -4 \\ 1 \end{bmatrix} \left(\frac{1}{42} \right) [5 \ -4 \ 1]$$

$$= \frac{1}{42} \begin{bmatrix} 25 & -20 & 5 \\ -20 & 16 & -4 \\ 5 & -4 & 1 \end{bmatrix}.$$

13. As in Exercise 9,

$$P_W = \begin{bmatrix} 1 \\ 2 \\ 1 \end{bmatrix} \left(\begin{bmatrix} 1 & 2 & 1 \end{bmatrix} \begin{bmatrix} 1 \\ 2 \\ 1 \end{bmatrix} \right)^{-1} \begin{bmatrix} 1 & 2 & 1 \end{bmatrix} = \frac{1}{6} \begin{bmatrix} 1 & 2 & 1 \\ 2 & 4 & 2 \\ 1 & 2 & 1 \end{bmatrix},$$

and hence the vector in W that is closest to $\mathbf{v}$ is

$$P_W \mathbf{v} = \frac{1}{6} \begin{bmatrix} 1 & 2 & 1 \\ 2 & 4 & 2 \\ 1 & 2 & 1 \end{bmatrix} \begin{bmatrix} -1 \\ 1 \\ 3 \end{bmatrix} = \frac{2}{3} \begin{bmatrix} 1 \\ 2 \\ 1 \end{bmatrix}.$$

The distance from $\mathbf{v}$ to W is

$$\| \mathbf{v} - P_W \mathbf{v} \| = \left\| \begin{bmatrix} -1 \\ 1 \\ 3 \end{bmatrix} - \frac{2}{3} \begin{bmatrix} 1 \\ 2 \\ 1 \end{bmatrix} \right\| = \frac{5\sqrt{3}}{3}.$$

17. Using the data, we let

$$C = \begin{bmatrix} 1 & 1 \\ 1 & 3 \\ 1 & 5 \\ 1 & 7 \end{bmatrix} \quad \text{and} \quad \mathbf{y} = \begin{bmatrix} 14 \\ 17 \\ 19 \\ 20 \end{bmatrix}.$$

Then the equation of the least squares line for the given data is $y = a_0 + a_1 x$, where

$$\begin{bmatrix} a_0 \\ a_1 \end{bmatrix} = (C^T C)^{-1} C^T \mathbf{y} = \begin{bmatrix} 4 & 16 \\ 16 & 84 \end{bmatrix}^{-1} \begin{bmatrix} 1 & 1 & 1 & 1 \\ 1 & 3 & 5 & 7 \end{bmatrix} \begin{bmatrix} 14 \\ 17 \\ 19 \\ 20 \end{bmatrix}$$

$$= \frac{1}{20} \begin{bmatrix} 21 & -4 \\ -4 & 1 \end{bmatrix} \begin{bmatrix} 70 \\ 300 \end{bmatrix} = \begin{bmatrix} 13.5 \\ 1.0 \end{bmatrix}.$$

Therefore $y = 13.5 + x$.

21. Using the data, we let

$$C = \begin{bmatrix} 1 & 1 \\ 1 & 3 \\ 1 & 7 \\ 1 & 8 \\ 1 & 10 \end{bmatrix} \quad \text{and} \quad \mathbf{y} = \begin{bmatrix} 40 \\ 36 \\ 23 \\ 21 \\ 13 \end{bmatrix}.$$

Then the equation of the least squares line for the given data is $y = a_0 + a_1 x$, where

$$\begin{bmatrix} a_0 \\ a_1 \end{bmatrix} = (C^T C)^{-1} C^T \mathbf{y} = \begin{bmatrix} 5 & 29 \\ 29 & 223 \end{bmatrix}^{-1} \begin{bmatrix} 1 & 1 & 1 & 1 & 1 \\ 1 & 3 & 7 & 8 & 10 \end{bmatrix} \begin{bmatrix} 40 \\ 36 \\ 23 \\ 21 \\ 13 \end{bmatrix}$$

$$= \frac{1}{274} \begin{bmatrix} 223 & -29 \\ -29 & 5 \end{bmatrix} \begin{bmatrix} 40 \\ 36 \\ 21 \\ 13 \end{bmatrix} = \begin{bmatrix} 44 \\ -3 \end{bmatrix}.$$

Therefore $y = 44 - 3x$.

25. Using the data, we let

$$C = \begin{bmatrix} 1 & -2 & 4 & -8 \\ 1 & -1 & 1 & -1 \\ 1 & 0 & 0 & 0 \\ 1 & 2 & 4 & 8 \\ 1 & 3 & 9 & 27 \end{bmatrix} \quad \text{and} \quad \mathbf{y} = \begin{bmatrix} -4 \\ 1 \\ 1 \\ 10 \\ 26 \end{bmatrix}.$$

Then the best cubic fit for the data is given by

$$y = a_0 + a_1 x + a_2 x^2 + a_3 x^3,$$

where

$$\begin{bmatrix} a_0 \\ a_1 \\ a_2 \\ a_3 \end{bmatrix} = (C^T C)^{-1} C^T \mathbf{y} = \begin{bmatrix} 5 & 2 & 18 & 26 \\ 2 & 18 & 26 & 114 \\ 18 & 26 & 114 & 242 \\ 26 & 114 & 242 & 858 \end{bmatrix}^{-1} \begin{bmatrix} 34 \\ 105 \\ 259 \\ 813 \end{bmatrix} \approx \begin{bmatrix} 1.42 \\ 0.49 \\ 0.38 \\ 0.73 \end{bmatrix}.$$

Therefore (using these approximate values), $y = 1.42 + 0.49x + 0.38x^2 + 0.73x^3$.

29. By Theorem 6.7 of Section 6.2, there are unique vectors $\mathbf{w}$ in W and $\mathbf{z}$ in $W^\perp$ such that $\mathbf{v} = \mathbf{w} + \mathbf{z}$. It follows that $\mathbf{v}$ is in $W^\perp$ if and only if $\mathbf{v} = \mathbf{z}$, and hence if and only if $\mathbf{w} = \mathbf{0}$. By Theorem 6.8, $P_W \mathbf{v} = \mathbf{w}$, and hence $P_W \mathbf{v} = \mathbf{0}$ if and only if $\mathbf{v}$ is in $W^\perp$.

ALTERNATE PROOF: By the boxed result on page 340, $P_W = C(C^T C)^{-1} C^T$, where C is a matrix whose columns are a basis for W. Now suppose that $\mathbf{v}$ is in $W^\perp$. Then $\mathbf{v}$ is orthogonal to each column of C, and hence $C^T \mathbf{v} = \mathbf{0}$. Therefore

$$P_W \mathbf{v} = C(C^T C)^{-1} C^T \mathbf{v} = C(C^T C)^{-1}(C^T \mathbf{v}) = C(C^T C)^{-1} \mathbf{0} = \mathbf{0}.$$

Conversely, suppose that $P_W \mathbf{v} = \mathbf{0}$. Then

$$C(C^T C)^{-1} C^T \mathbf{v} = \mathbf{0}$$
$$C^T C(C^T C)^{-1} C^T \mathbf{v} = C^T \mathbf{0} = \mathbf{0}$$
$$C^T \mathbf{v} = \mathbf{0}.$$

This last equation asserts that $\mathbf{v}$ is orthogonal to the columns of C, a spanning set for W. Therefore $\mathbf{v}$ is in $W^\perp$ by Exercise 31 of Section 6.2.

33. Let $\mathbf{v}$ be a vector in $\mathcal{R}^n$, and let $\mathbf{w}$ and $\mathbf{z}$ be the unique vectors in W and $W^\perp$, respectively, such that $\mathbf{v} = \mathbf{w} + \mathbf{z}$. Then

$$(P_W + P_{W^\perp})\mathbf{v} = P_W\mathbf{v} + P_{W^\perp}\mathbf{v} = \mathbf{w} + \mathbf{z} = \mathbf{v} = I_n\mathbf{v}.$$

Since $\mathbf{v}$ is an arbitrarily chosen vector in $\mathcal{R}^n$, the result follows.

37. (a) We first show that 1 and 0 are eigenvalues of P_W. Since $k \neq 0$, we can choose a nonzero vector $\mathbf{w}$ in W. Then $P_W\mathbf{w} = \mathbf{w}$, and hence $\mathbf{w}$ is an eigenvector with corresponding eigenvalue 1. Since $k \neq n$, we can choose a nonzero vector $\mathbf{z}$ in $W^\perp$. Then $P_W\mathbf{z} = \mathbf{0}$, and hence $\mathbf{z}$ is an eigenvector with corresponding eigenvalue 0.

Next we show that 1 and 0 are the only eigenvalues of P_W. Suppose that λ is a nonzero eigenvalue of P_W, and let $\mathbf{v}$ be an eigenvector corresponding to λ. Then $P_W\mathbf{v} = \lambda\mathbf{v}$, and hence $P_W(\frac{1}{\lambda}\mathbf{v}) = \mathbf{v}$. It follows that $\mathbf{v}$ is in W, and hence $P_W\mathbf{v} = \mathbf{v}$. Therefore $\lambda = 1$.

(b) Since $P_W\mathbf{v} = \mathbf{v}$ if and only $\mathbf{v}$ is in W, we see that W is the eigenspace of P_W corresponding to eigenvalue 1. Similarly, since $P_W\mathbf{v} = \mathbf{0}$ if and only if $\mathbf{v}$ is in $W^\perp$, we have that $W^\perp$ is the eigenspace of P_W corresponding to eigenvalue 0.

(c) Let T be the matrix transformation induced by P_W. Since $T(\mathbf{v}) = 1 \cdot \mathbf{v}$ for all $\mathbf{v}$ in $\mathcal{B}_1$ and $T(\mathbf{v}) = 0 \cdot \mathbf{v}$ for all $\mathbf{v}$ in $\mathcal{B}_2$, we have $[T]_\mathcal{B} = D$, and hence $P_W = BDB^{-1}$ by Theorem 4.10.

41. Since $\operatorname{Col} A = \operatorname{Row} A^T$, we can obtain a basis for $\operatorname{Col} A$ by choosing a basis for $\operatorname{Row} A^T$. The reduced row echelon form of A^T is

$$R = \begin{bmatrix} 1 & 0 & -1 & 1 \\ 0 & 1 & -2 & 1 \\ 0 & 0 & 0 & 0 \\ 0 & 0 & 0 & 0 \end{bmatrix}.$$

Since the nonzero rows of R form a basis for $\operatorname{Row} A^T$, these same rows, written as column vectors,

$$\mathcal{B}_1 = \left\{ \begin{bmatrix} 1 \\ 0 \\ -1 \\ 1 \end{bmatrix}, \begin{bmatrix} 0 \\ 1 \\ -2 \\ 1 \end{bmatrix} \right\}$$

form a basis for $\operatorname{Col} A$. Next observe that $(\operatorname{Col} A)^\perp$ is the solution space to the system $A^T\mathbf{x} = \mathbf{0}$. We may use R, the reduced row echelon form of A^T, to obtain the general solution and a basis for this solution space. The resulting basis for $(\operatorname{Col} A)^\perp$ is:

$$\mathcal{B}_2 = \left\{ \begin{bmatrix} 1 \\ 2 \\ 1 \\ 0 \end{bmatrix}, \begin{bmatrix} -1 \\ -1 \\ 0 \\ 1 \end{bmatrix} \right\}.$$

Let

$$
B = \begin{bmatrix} 1 & 0 & 1 & -1 \\ 0 & 1 & 2 & -1 \\ -1 & -2 & 1 & 0 \\ 1 & 1 & 0 & 1 \end{bmatrix} \quad \text{and} \quad D = \begin{bmatrix} 1 & 0 & 0 & 0 \\ 0 & 1 & 0 & 0 \\ 0 & 0 & 0 & 0 \\ 0 & 0 & 0 & 0 \end{bmatrix}.
$$

Notice that B is the matrix whose columns are the vectors in $\mathcal{B}_1 \cup \mathcal{B}_2$. Then by Exercise 37,

$$
P_W = BDB^{-1} = \begin{bmatrix} 1 & 0 & 1 & -1 \\ 0 & 1 & 2 & -1 \\ -1 & -2 & 1 & 0 \\ 1 & 1 & 0 & 1 \end{bmatrix} \begin{bmatrix} 1 & 0 & 0 & 0 \\ 0 & 1 & 0 & 0 \\ 0 & 0 & 0 & 0 \\ 0 & 0 & 0 & 0 \end{bmatrix} \frac{1}{3} \begin{bmatrix} 2 & -1 & 0 & 1 \\ -1 & 1 & -1 & 0 \\ 0 & 1 & 1 & 1 \\ -1 & 0 & 1 & 2 \end{bmatrix}
$$

$$
= \frac{1}{3} \begin{bmatrix} 2 & -1 & 0 & 1 \\ -1 & 1 & -1 & 0 \\ 0 & -1 & 2 & -1 \\ 1 & 0 & -1 & 1 \end{bmatrix}.
$$

45. (rounded to 4 places after the decimal)

$$
P_W = \begin{bmatrix} 0.7201 & 0.0001 & -0.1845 & -0.3943 & -0.1098 \\ 0.0001 & 0.4915 & 0.4391 & -0.1547 & -0.1823 \\ -0.1845 & 0.4391 & 0.4993 & -0.1263 & 0.0850 \\ -0.3943 & -0.1547 & -0.1263 & 0.3975 & -0.2102 \\ -0.1098 & -0.1823 & 0.0850 & -0.2102 & 0.8915 \end{bmatrix}
$$

The distance equals 3.4418.

6.4   ORTHOGONAL MATRICES AND OPERATORS

1. (a) True

(b) False; for example, if T is a translation by a nonzero vector, then T preserves distances, but T is not linear.

(c) False, only orthogonal linear operators preserve dot products.

(d) True

(e) True

(f) True

(g) False, for example, let $P = I_n$ and $Q = -I_n$.

(h) False, for example, let $P = \begin{bmatrix} 1 & 1 \\ 1 & 2 \end{bmatrix}$.

5. Since

$$\begin{bmatrix} 0 & 1 & 0 \\ 0 & 0 & 1 \\ 1 & 0 & 0 \end{bmatrix} \begin{bmatrix} 0 & 1 & 0 \\ 0 & 0 & 1 \\ 1 & 0 & 0 \end{bmatrix}^T = \begin{bmatrix} 0 & 1 & 0 \\ 0 & 0 & 1 \\ 1 & 0 & 0 \end{bmatrix} \begin{bmatrix} 0 & 0 & 1 \\ 1 & 0 & 0 \\ 0 & 1 & 0 \end{bmatrix} = I_3,$$

the matrix is invertible, and its inverse is equal to its transpose. Therefore the matrix is orthogonal by Theorem 6.9(b).

7. No, the matrix is not a square matrix.

9. Since $\det\left(\dfrac{1}{\sqrt{2}}\begin{bmatrix} 1 & 1 \\ 1 & -1 \end{bmatrix}\right) = \frac{1}{2}(-1-1) = -1$, the operator is a reflection. The line of reflection is the same as the eigenspace of the matrix corresponding to eigenvalue 1. This is the solution space to the system

$$\left(\frac{1}{\sqrt{2}}\begin{bmatrix} 1 & 1 \\ 1 & -1 \end{bmatrix} - \begin{bmatrix} 1 & 0 \\ 0 & 1 \end{bmatrix}\right)\begin{bmatrix} x \\ y \end{bmatrix} = \begin{bmatrix} 0 \\ 0 \end{bmatrix},$$

or

$$(\tfrac{1}{\sqrt{2}} - 1)x + \tfrac{1}{\sqrt{2}}y = 0$$
$$\tfrac{1}{\sqrt{2}}x - \tfrac{1}{\sqrt{2}}y = 0.$$

Solving this system for y in terms of x, we obtain $y = (\sqrt{2}-1)x$. This equation is the line of reflection.

11. Since $\det\left(\dfrac{1}{2}\begin{bmatrix} \sqrt{3} & -1 \\ 1 & \sqrt{3} \end{bmatrix}\right) = \frac{1}{4}(3+1) = 1$, the operator is a rotation. Comparing this matrix with the general form of a rotation matrix,

$$A_\theta = \begin{bmatrix} \cos\theta & -\sin\theta \\ \sin\theta & \cos\theta \end{bmatrix} = \begin{bmatrix} \frac{\sqrt{3}}{2} & -\frac{1}{2} \\ \frac{1}{2} & \frac{\sqrt{3}}{2} \end{bmatrix},$$

we see that $\cos\theta = \frac{\sqrt{3}}{2}$ and $\sin\theta = \frac{1}{2}$. Therefore $\theta = 30°$.

13. Since $\det\left(\dfrac{1}{13}\begin{bmatrix} 5 & 12 \\ 12 & -5 \end{bmatrix}\right) = \frac{1}{13^2}(-25-144) = -1$, the operator is a reflection. As in Exercise 9, the line of reflection is the eigenspace of the given matrix corresponding to eigenvalue 1, which is the solution space to the system

$$\left(\frac{1}{13}\begin{bmatrix} 5 & 12 \\ 12 & -5 \end{bmatrix} - \begin{bmatrix} 1 & 0 \\ 0 & 1 \end{bmatrix}\right)\begin{bmatrix} x \\ y \end{bmatrix} = \begin{bmatrix} 0 \\ 0 \end{bmatrix}.$$

Solving this system for y in terms of x, we obtain $y = \frac{2}{3}x$. This equation is the line of reflection.

17. **(a)** Let Q be the standard matrix of T. Then

$$Q = \begin{bmatrix} \cos\theta & \sin\theta & 0 \\ \sin\theta & \cos\theta & 0 \\ 0 & 0 & 1 \end{bmatrix}.$$

It is a simple matter to verify that $QQ^T = I_3$, showing that Q is an orthogonal matrix. It follows that T is an orthogonal operator.

(b) First, we compute the characteristic polynomial of Q in (a):

$$\det(Q - tI_3) = \det \begin{bmatrix} \cos\theta - t & \sin\theta & 0 \\ \sin\theta & \cos\theta - t & 0 \\ 0 & 0 & 1-t \end{bmatrix} = (1-t)(t^2 - 2t\cos\theta + 1).$$

Since the discriminant of $t^2 - 2t\cos\theta + 1$ is $4\cos^2\theta - 4 < 0$, the only (real) eigenvalue of Q is 1, which has multiplicity 1. It follows that the eigenspace corresponding to 1 is 1-dimensional. Since $T(\mathbf{e}_3) = \mathbf{e}_3$, we see that $\mathbf{e}_3$ is an eigenvector with eigenvalue 1. Therefore the eigenspace corresponding to 1 is the space spanned by the set $\{\mathbf{e}_3\}$, the z-axis.

(c) For any vector $\mathbf{v}$ in $\mathcal{R}^3$, the image $T(\mathbf{v})$ is obtained by rotating $\mathbf{v}$ by the angle θ about the z-axis. Looking down from the positive direction of the z-axis, we see that this rotation is counter-clockwise.

21. Let $\mathcal{B} = \{\mathbf{v}, \mathbf{w}\}$. Observe that

$$T(\mathbf{v}) = (\mathbf{v} \cdot \mathbf{v} \cos\theta + \mathbf{v} \cdot \mathbf{w} \sin\theta)\mathbf{v} + (-\mathbf{v} \cdot \mathbf{v} \sin\theta + \mathbf{v} \cdot \mathbf{w} \cos\theta)\mathbf{w}$$
$$= \cos\theta\mathbf{v} - \sin\theta\mathbf{w}.$$

Similarly, $T(\mathbf{w}) = \sin\theta\mathbf{v} + \cos\theta\mathbf{w}$, and hence

$$[T]_\mathcal{B} = \begin{bmatrix} \cos\theta & \sin\theta \\ -\sin\theta & \cos\theta \end{bmatrix}.$$

It is a simple matter to verify that $[T]_\mathcal{B}$ is an orthogonal matrix, and therefore T is an orthogonal operator.

25. Suppose that λ is an eigenvalue for Q, and let $\mathbf{v}$ be a corresponding eigenvector. Then

$$\|\mathbf{v}\| = \|Q\mathbf{v}\| = \|\lambda\mathbf{v}\| = |\lambda|\|\mathbf{v}\|.$$

Since $\|\mathbf{v}\| \neq 0$, it follows that $|\lambda| = 1$. Therefore $\lambda = \pm 1$.

29. **(a)** $Q_W^T = (2P_W - I_2)^T = 2P_W^T - I_2^T = 2P_W - I_2 = Q_W$

(b) $Q_W^2 = (2P_W - I_2)^2 = 4P_W^2 - 4P_W I_2 + I_2 = 4P_W - 4P_W + I_2 = I_2$

(c) By (a) and (b), we have $Q_W Q_W^T = Q_W Q_W = I_2$, and hence Q_W is an orthogonal matrix.

(d) $Q_W \mathbf{w} = (2P_W - I_2)\mathbf{w} = 2P_W \mathbf{w} - I_2 \mathbf{w} = 2\mathbf{w} - \mathbf{w} = \mathbf{w}$

(e) $Q_W \mathbf{v} = (2P_W - I_2)\mathbf{v} = 2P_W \mathbf{v} - I_2 \mathbf{v} = \mathbf{0} - \mathbf{v} = -\mathbf{v}$

(f) Select nonzero vectors $\mathbf{w}$ in W and $\mathbf{v}$ in $W^{\perp}$. Then $\{\mathbf{w}, \mathbf{v}\}$ is a basis for $\mathcal{R}^2$ since the set is an orthogonal set of nonzero vectors. Let $P = [\mathbf{w} \ \mathbf{v}]$, and let T be the matrix transformation induced by Q_W. Then Q_W is the standard matrix of T, and T is an orthogonal operator because Q_W is an orthogonal matrix. Also,

$$Q_W = PDP^{-1}, \qquad \text{where} \qquad D = \begin{bmatrix} 1 & 0 \\ 0 & -1 \end{bmatrix}.$$

Thus

$$\begin{aligned}
\det Q_W &= \det (PDP^{-1}) \\
&= (\det P)(\det D)(\det P^{-1}) \\
&= (\det P)(-1)(\det P)^{-1} = -1.
\end{aligned}$$

It follows that T is a reflection. Furthermore, since $T(\mathbf{w}) = \mathbf{w}$, T is the reflection of $\mathcal{R}^2$ about W.

33. $\|T(\mathbf{u})\| = \|T(\mathbf{u}) - \mathbf{0}\| = \|T(\mathbf{u}) - T(\mathbf{0})\| = \|\mathbf{u} - \mathbf{0}\| = \|\mathbf{u}\|$

37. Since

$$F\left(\begin{bmatrix} 1 \\ 0 \end{bmatrix}\right) + F\left(\begin{bmatrix} 0 \\ 1 \end{bmatrix}\right) = Q\begin{bmatrix} 1 \\ 0 \end{bmatrix} + \mathbf{b} + Q\begin{bmatrix} 0 \\ 1 \end{bmatrix} + \mathbf{b}$$
$$= \mathbf{q}_1 + \mathbf{q}_2 + 2\mathbf{b}$$

and

$$F\left(\begin{bmatrix} 1 \\ 1 \end{bmatrix}\right) = Q\left(\begin{bmatrix} 1 \\ 0 \end{bmatrix} + \begin{bmatrix} 0 \\ 1 \end{bmatrix}\right) + \mathbf{b}$$
$$= \mathbf{q}_1 + \mathbf{q}_2 + \mathbf{b},$$

it follows that

$$\mathbf{b} = F\left(\begin{bmatrix} 1 \\ 0 \end{bmatrix}\right) + F\left(\begin{bmatrix} 0 \\ 1 \end{bmatrix}\right) - F\left(\begin{bmatrix} 1 \\ 1 \end{bmatrix}\right)$$
$$= \begin{bmatrix} 2 \\ 4 \end{bmatrix} + \begin{bmatrix} 1 \\ 3 \end{bmatrix} - \begin{bmatrix} 2 \\ 3 \end{bmatrix} = \begin{bmatrix} 1 \\ 4 \end{bmatrix}.$$

Thus

$$\mathbf{q}_1 = Q\begin{bmatrix} 1 \\ 0 \end{bmatrix} = F\left(\begin{bmatrix} 1 \\ 0 \end{bmatrix}\right) - \mathbf{b} = \begin{bmatrix} 2 \\ 4 \end{bmatrix} - \begin{bmatrix} 1 \\ 4 \end{bmatrix} = \begin{bmatrix} 1 \\ 0 \end{bmatrix}$$

and

$$\mathbf{q}_2 = Q \begin{bmatrix} 0 \\ 1 \end{bmatrix} = F \left(\begin{bmatrix} 0 \\ 1 \end{bmatrix} \right) - \mathbf{b} = \begin{bmatrix} 1 \\ 3 \end{bmatrix} - \begin{bmatrix} 1 \\ 4 \end{bmatrix} = \begin{bmatrix} 0 \\ -1 \end{bmatrix}.$$

Therefore $Q = [\mathbf{q}_1 \ \mathbf{q}_2] = \begin{bmatrix} 1 & 0 \\ 0 & -1 \end{bmatrix}$.

41. (a) We have

$$A_2 = \begin{bmatrix} 0 & -1 \\ -1 & 0 \end{bmatrix},$$

$$A_3 = I_3 - \frac{2}{3} E_3 = \begin{bmatrix} 1 & 0 & 0 \\ 0 & 1 & 0 \\ 0 & 0 & 1 \end{bmatrix} - \frac{2}{3} \begin{bmatrix} 1 & 1 & 1 \\ 1 & 1 & 1 \\ 1 & 1 & 1 \end{bmatrix} = \frac{1}{3} \begin{bmatrix} 1 & -2 & -2 \\ -2 & 1 & -2 \\ -2 & -2 & 1 \end{bmatrix},$$

and

$$A_6 = \frac{1}{3} \begin{bmatrix} 2 & -1 & -1 & -1 & -1 & -1 \\ -1 & 2 & -1 & -1 & -1 & -1 \\ -1 & -1 & 2 & -1 & -1 & -1 \\ -1 & -1 & -1 & 2 & -1 & -1 \\ -1 & -1 & -1 & -1 & 2 & -1 \\ -1 & -1 & -1 & -1 & -1 & 2 \end{bmatrix}.$$

(b) We verify the conclusion for the case $n = 3$.

$$A_3^T A_3 = \frac{1}{3} \begin{bmatrix} 1 & -2 & -2 \\ -2 & 1 & -2 \\ -2 & -2 & 1 \end{bmatrix} \frac{1}{3} \begin{bmatrix} 1 & -2 & -2 \\ -2 & 1 & -2 \\ -2 & -2 & 1 \end{bmatrix}$$

$$= \frac{1}{9} \begin{bmatrix} 9 & 0 & 0 \\ 0 & 9 & 0 \\ 0 & 0 & 9 \end{bmatrix} = I_3.$$

It follows that A_3 is invertible and $(A_3)^{-1} = A_3^T$. Therefore A_3 is orthogonal by Theorem 6.2(b).

(c) Since E_n is a square matrix with each entry equal to 1, we have $E_n^T = E_n$. Thus

$$A_n^T = (I_n - \frac{2}{n} E_n)^T = I_n^T - \frac{2}{n} E_n^T = I_n - \frac{2}{n} E_n = A_n,$$

proving that A_n is symmetric.

(d) First observe that, for every i and j, the (i,j)-entry of E_n^2 is given by

$$1^2 + 1^2 + \cdots + 1^2 = n.$$

Hence $E_n^2 = nE_n$. Therefore

$$\begin{aligned}
A_n^2 &= (I_n - \frac{2}{n}E_n)(I_n - \frac{2}{n}E_n) \\
&= I_n - \frac{4}{n}I_nE_n + \frac{4}{n^2}e_n^2 \\
&= I_n - \frac{4}{n}E_n + \frac{4}{n^2}nE_n \\
&= I_n.
\end{aligned}$$

Thus, by (c), we have that $A_nA_n^T = A_nA_n = I_n$. It follows that A_n is invertible and $(A_n)^{-1} = A_n^T$. Therefore A_n is orthogonal by Theorem 6.2(b).

45. $\begin{bmatrix} 0.7833 & 0.6217 \\ 0.6217 & -0.7833 \end{bmatrix}$ (rounded to 4 places after the decimal)

6.5 SYMMETRIC MATRICES

1. (a) True
   (b) False, any nonzero vector in $\mathcal{R}^2$ is an eigenvector of the symmetric matrix I_2, but not every 2×2 with nonzero columns is an orthogonal matrix.
   (c) False, let $A = \begin{bmatrix} 1 & 4 \\ 1 & 1 \end{bmatrix}$. Then $\begin{bmatrix} 2 \\ 1 \end{bmatrix}$ and $\begin{bmatrix} 2 \\ -1 \end{bmatrix}$ are eigenvectors of A that correspond to the eigenvalues 3 and -1, respectively. However, they are not orthogonal.
   (d) False, if $\mathbf{v}$ is an eigenvector, then so is $2\mathbf{v}$. But these two eigenvectors are not orthogonal.
   (e) True
   (f) True
   (g) False, $\begin{bmatrix} 1 & 0 \\ 0 & -1 \end{bmatrix}$ is not the sum of orthogonal projection matrices.
   (h) True

5. (a) Let $a_{11} = 5$, $a_{22} = 5$, and $a_{12} = a_{21} = \frac{4}{2} = 2$. Then

$$A = \begin{bmatrix} 5 & 2 \\ 2 & 5 \end{bmatrix}.$$

The eigenvalues of A are 7 and 3, and

$$\left\{ \begin{bmatrix} \frac{1}{\sqrt{2}} \\ \frac{1}{\sqrt{2}} \end{bmatrix}, \begin{bmatrix} -\frac{1}{\sqrt{2}} \\ \frac{1}{\sqrt{2}} \end{bmatrix} \right\}$$

is an orthonormal basis for $\mathcal{R}^2$ consisting of corresponding eigenvectors. Choose the x'-axis to be in the direction of the first vector in this basis.

(b)   Since the x'-axis is in the direction of $\begin{bmatrix} \frac{1}{\sqrt{2}} \\ \frac{1}{\sqrt{2}} \end{bmatrix}$, it has a slope of 1, and hence the

angle of rotation of the coordinate system is $45°$.

(c)   Let P be the matrix whose columns are the vectors in the orthonormal basis given in (a). Then

$$P = \begin{bmatrix} \frac{1}{\sqrt{2}} & -\frac{1}{\sqrt{2}} \\ \frac{1}{\sqrt{2}} & \frac{1}{\sqrt{2}} \end{bmatrix}, \qquad \text{and} \qquad \begin{bmatrix} x \\ y \end{bmatrix} = P \begin{bmatrix} x' \\ y' \end{bmatrix}.$$

Therefore

$$x = \tfrac{1}{\sqrt{2}}x' - \tfrac{1}{\sqrt{2}}y'$$
$$y = \tfrac{1}{\sqrt{2}}x' + \tfrac{1}{\sqrt{2}}y'.$$

(d)   Since the eigenvalues are 7 and 3, the equation becomes $7(x')^2 + 3(y')^2 = 9$.

(e)   The conic section is an ellipse.

9.   (a)   Let $a_{11} = 2$, $a_{22} = -7$, and $a_{12} = a_{21} = \frac{-12}{2} = -6$. Then

$$A = \begin{bmatrix} 2 & -6 \\ -6 & -7 \end{bmatrix}.$$

The eigenvalues of A are -10 and 5, and

$$\left\{ \begin{bmatrix} \frac{1}{\sqrt{5}} \\ \frac{2}{\sqrt{5}} \end{bmatrix}, \begin{bmatrix} -\frac{2}{\sqrt{5}} \\ \frac{1}{\sqrt{5}} \end{bmatrix} \right\}$$

is an orthonormal basis for $\mathcal{R}^2$ consisting of corresponding eigenvectors. Choose the x'-axis to be in the direction of the first vector in this basis.

(b)   Since the x'-axis is in the direction of $\begin{bmatrix} \frac{1}{\sqrt{5}} \\ \frac{2}{\sqrt{5}} \end{bmatrix}$, it has a slope of 2, and hence the

angle of rotation of the coordinate system is $\tan^{-1} 2 \approx 63.4°$.

(c)   Let P be the matrix whose columns are the vectors in the orthonormal basis given in (a). Then

$$P = \begin{bmatrix} \frac{1}{\sqrt{5}} & -\frac{2}{\sqrt{5}} \\ \frac{2}{\sqrt{5}} & \frac{1}{\sqrt{5}} \end{bmatrix} \qquad \text{and} \qquad \begin{bmatrix} x \\ y \end{bmatrix} = P \begin{bmatrix} x' \\ y' \end{bmatrix}.$$

Therefore

$$x = \tfrac{1}{\sqrt{5}}x' - \tfrac{2}{\sqrt{5}}y'$$
$$y = \tfrac{2}{\sqrt{5}}x' + \tfrac{1}{\sqrt{5}}y'.$$

(d) Since the eigenvalues are -10 and 5, the equation becomes $-10(x')^2 + 5(y')^2 = 200$.

(e) The conic section is a hyperbola.

13. The characteristic polynomial of A is

$$\det(A - tI_2) = \det \begin{bmatrix} t-3 & 1 \\ 1 & t-3 \end{bmatrix} = (t-3)^2 - 1 = (t-2)(t-4),$$

and hence A has the eigenvalues $\lambda_1 = 2$ and $\lambda_2 = 4$. The vectors $\begin{bmatrix} 1 \\ 1 \end{bmatrix}$ and $\begin{bmatrix} 1 \\ -1 \end{bmatrix}$ are eigenvectors that correspond to these eigenvalues. Normalizing these vectors, we obtain unit vectors

$$\mathbf{u}_1 = \begin{bmatrix} \frac{1}{\sqrt{2}} \\ -\frac{1}{\sqrt{2}} \end{bmatrix} \quad \text{and} \quad \mathbf{u}_2 = \begin{bmatrix} \frac{1}{\sqrt{2}} \\ \frac{1}{\sqrt{2}} \end{bmatrix},$$

which constitute an orthonormal basis $\{\mathbf{u}_1, \mathbf{u}_2\}$ for $\mathcal{R}^2$. We use these eigenvectors and corresponding eigenvalues to obtain the spectral decomposition

$$A = \lambda_1 \mathbf{u}_1 \mathbf{u}_1^T + \lambda_2 \mathbf{u}_2 \mathbf{u}_2^T$$

$$= 2 \begin{bmatrix} \frac{1}{\sqrt{2}} \\ -\frac{1}{\sqrt{2}} \end{bmatrix} \begin{bmatrix} \frac{1}{\sqrt{2}} & -\frac{1}{\sqrt{2}} \end{bmatrix} + 4 \begin{bmatrix} \frac{1}{\sqrt{2}} \\ \frac{1}{\sqrt{2}} \end{bmatrix} \begin{bmatrix} \frac{1}{\sqrt{2}} & \frac{1}{\sqrt{2}} \end{bmatrix}$$

$$= 2 \begin{bmatrix} 0.5 & -0.5 \\ -0.5 & 0.5 \end{bmatrix} + 4 \begin{bmatrix} 0.5 & 0.5 \\ 0.5 & 0.5 \end{bmatrix}.$$

17. The characteristic polynomial of A is

$$\det(A - tI_3) = \det \begin{bmatrix} 3-t & 2 & 2 \\ 2 & 2-7 & 0 \\ 2 & 0 & 4-t \end{bmatrix} = -18t + 9t^2 - t^3 = -t(t-3)(t-6),$$

and hence A has the eigenvalues $\lambda_1 = 3$, $\lambda_2 = 6$, and $\lambda_3 = 0$. For each λ_i, select a nonzero solution to $(A - \lambda_i I_3)\mathbf{x} = \mathbf{0}$ to obtain an eigenvector corresponding to each eigenvalue. Since the eigenvalues are distinct, the eigenvectors are orthogonal, and hence normalizing these eigenvectors results in an orthonormal basis for $\mathcal{R}^3$ consisting of eigenvectors of A. One example of such normalized eigenvectors is

$$\mathbf{u}_1 = \frac{1}{3} \begin{bmatrix} -1 \\ -2 \\ 2 \end{bmatrix}, \quad \mathbf{u}_2 = \frac{1}{3} \begin{bmatrix} 2 \\ 1 \\ 2 \end{bmatrix}, \quad \text{and} \quad \mathbf{u}_3 = \frac{1}{3} \begin{bmatrix} -2 \\ 2 \\ 1 \end{bmatrix}.$$

Thus $\{\mathbf{u}_1, \mathbf{u}_2, \mathbf{u}_3\}$ is an orthonormal basis for $\mathcal{R}^3$ consisting of eigenvectors of A. We use these eigenvectors and corresponding eigenvalues to obtain the spectral decomposition

$$A = \lambda_1 \mathbf{u}_1 \mathbf{u}_1^T + \lambda_2 \mathbf{u}_2 \mathbf{u}_2^T + \lambda_3 \mathbf{u}_3 \mathbf{u}_3^T$$

$$= 3\left(\frac{1}{3}\right)\begin{bmatrix} -1 \\ -2 \\ 2 \end{bmatrix}\frac{1}{3}[-1 \ \ -2 \ \ 2] + 6\left(\frac{1}{3}\right)\begin{bmatrix} 2 \\ 1 \\ 2 \end{bmatrix}\frac{1}{3}[2 \ \ 1 \ \ 2] + 0\left(\frac{1}{3}\right)\begin{bmatrix} -2 \\ 2 \\ 1 \end{bmatrix}\frac{1}{3}[-2 \ \ 2 \ \ 1]$$

$$= 3\begin{bmatrix} \frac{1}{9} & \frac{2}{9} & -\frac{2}{9} \\ \frac{2}{9} & \frac{4}{9} & -\frac{4}{9} \\ -\frac{2}{9} & -\frac{4}{9} & \frac{2}{9} \end{bmatrix} + 6\begin{bmatrix} \frac{4}{9} & \frac{2}{9} & \frac{4}{9} \\ \frac{2}{9} & \frac{1}{9} & \frac{2}{9} \\ \frac{4}{9} & \frac{2}{9} & \frac{4}{9} \end{bmatrix} + 0\begin{bmatrix} \frac{4}{9} & -\frac{4}{9} & -\frac{2}{9} \\ -\frac{4}{9} & \frac{4}{9} & \frac{2}{9} \\ -\frac{2}{9} & \frac{2}{9} & \frac{1}{9} \end{bmatrix}.$$

21. $2\begin{bmatrix} 1 & 0 \\ 0 & 0 \end{bmatrix} + 2\begin{bmatrix} 0 & 0 \\ 0 & 1 \end{bmatrix}$ and $2\begin{bmatrix} .5 & .5 \\ .5 & .5 \end{bmatrix} + 2\begin{bmatrix} .5 & -.5 \\ -.5 & .5 \end{bmatrix}$

25. For $i = j$, we have

$$P_i P_i = \mathbf{u}_i \mathbf{u}_i^T \mathbf{u}_i \mathbf{u}_i^T = \mathbf{u}_i(\mathbf{u}_i^T \mathbf{u}_i)\mathbf{u}_i^T = \mathbf{u}_i(1)\mathbf{u}^T = \mathbf{u}_i \mathbf{u}_i^T = P_i;$$

and for $i \neq j$, we have

$$P_i P_j = \mathbf{u}_i \mathbf{u}_i^T \mathbf{u}_j \mathbf{u}_j^T = \mathbf{u}_i(\mathbf{u}_i^T \mathbf{u}_j)\mathbf{u}_j^T = \mathbf{u}_i[0]\mathbf{u}_j^T = O.$$

29. Suppose that $Q_j = P_r + P_{r+1} + \cdots + P_s$. Then

$$Q_j^T = (P_r + P_{r+1} + \cdots + P_s)^T = P_r^T + P_{r+1}^T + \cdots + P_s^T = P_r + P_{r+1} + \cdots + P_s = Q_j.$$

33. Let $A = \mu_1 Q_1 + \mu_2 Q_2 + \cdots + \mu_k Q_k$ be the spectral decomposition, as in Exercise 27. Then A^s is the sum of all products of s terms (with possible duplication) from the sum above. Any such term containing factors Q_i and Q_j with $i \neq j$ equals O. Otherwise, each factor of the term is of the form $\mu_i Q_i$, and hence the nonzero terms are of the form $\mu_i^s Q_i^s = \mu_i^s Q_i$. Therefore

$$A^s = \mu_1^s Q_1 + \mu_2^s Q_2 + \cdots + \mu_n^s Q_n.$$

37. Let $A = \mu_1 Q_1 + \mu_2 Q_2 + \cdots + \mu_k Q_k$ be the spectral decomposition, as in Exercise 27. By Exercise 35, we have

$$f_j(A) = f_j(\mu_1)Q_1 + \cdots + f_j(\mu_j)Q_j + \cdots + f_k(\mu_j)Q_k$$
$$= 0Q_1 + \cdots + 1Q_j + \cdots + 0Q_k = Q_j.$$

41. Suppose that A is positive definite. Then A is symmetric, and hence A^{-1} is also symmetric. Furthermore, the eigenvalues of A^{-1} are the reciprocals of the eigenvalues of A. Therefore, since the eigenvalues of A are positive, the eigenvalues of A^{-1} are also positive. It follows that A^{-1} is positive definite by Exercise 39.

45. If A and B are positive semidefinite $n \times n$ matrices, then $A + B$ is positive semidefinite.

First note that $A + B$ is symmetric because both A and B are symmetric. Let $\mathbf{v}$ be a nonzero vector in $\mathcal{R}^n$. Then $\mathbf{v}^T A \mathbf{v} \geq 0$ and $\mathbf{v}^T B \mathbf{v} \geq 0$ because A and B are positive semidefinite. Therefore

$$\mathbf{v}^T (A + B)\mathbf{v} = \mathbf{v}^T A \mathbf{v} + \mathbf{v}^T B \mathbf{v} \geq 0,$$

and hence $A + B$ is positive semidefinite.

49. If A is a positive semidefinite matrix, then there exists a positive semidefinite matrix B such that $B^2 = A$.

Suppose that $\mu_1, \mu_2, \ldots, \mu_k$ are the distinct eigenvalues of A, and that

$$A = \mu_1 Q_1 + \mu_2 Q_2 + \cdots + \mu_k Q_k$$

is a spectral decomposition of A as given in Exercise 27. Since each μ_i is nonnegative, it has a unique nonnegative square root, ν_i. Let

$$B = \nu_1 Q_1 + \nu_2 Q_2 + \cdots + \nu_k Q_k.$$

Then B is positive semidefinite because the eigenvalues of B, the ν_i, are nonnegative. Furthermore, by Exercise 33,

$$B^2 = \nu_1^2 Q_1 + \nu_2^2 Q_2 + \cdots + \nu_k^2 Q_k = \mu_1 Q_1 + \mu_2 Q_2 + \cdots + \mu_k Q_k = A.$$

53. By the argument on page 90, $A^T A$ and $A A^T$ are symmetric. Let $\mathbf{v}$ be any nonzero vector in $\mathcal{R}^n$. Since A is invertible, $A\mathbf{v} \neq \mathbf{0}$ by Theorem 2.6 on page 126. Thus

$$\mathbf{v}^T A^T A \mathbf{v} = (A\mathbf{v})^T (A\mathbf{v}) = (A\mathbf{v}) \cdot (A\mathbf{v}) > 0,$$

and therefore $A^T A$ is positive definite. Similarly, $A A^T$ is positive definite.

6.6   SINGULAR VALUE DECOMPOSITION

1. (a) False, σ^2 is an eigenvalue of $A^T A$.
  (b) True
  (c) False, see Example 7.
  (d) True
  (e) False, every matrix has a pseudoinverse.

5. First, compute

$$A^T A = \begin{bmatrix} 1 & 1 & 1 \\ 1 & -1 & 2 \end{bmatrix} \begin{bmatrix} 1 & 1 \\ 1 & -1 \\ 1 & 2 \end{bmatrix} = \begin{bmatrix} 3 & 2 \\ 2 & 6 \end{bmatrix}.$$

Second, determine the eigenvalues of $A^T A$ from the characteristic polynomial

$$\det\left(A^T A - tI_2\right) = \det \begin{bmatrix} 3-t & 2 \\ 2 & 6-t \end{bmatrix} = (t-7)(t-2).$$

Thus $\lambda_1 = 7$ and $\lambda_2 = 2$ are the eigenvalues of $A^T A$. The singular values of A are therefore $\sigma_1 = \sqrt{7}$ and $\sigma_2 = \sqrt{2}$. For each eigenvalue λ_i, find an eigenvector corresponding to λ_i by choosing a nonzero solution to $(A - \lambda_i I_2)\mathbf{x} = \mathbf{0}$. For example, choose

$$\mathbf{w}_1 = \begin{bmatrix} 1 \\ 2 \end{bmatrix} \quad \text{and} \quad \mathbf{w}_2 = \begin{bmatrix} 2 \\ -1 \end{bmatrix}$$

as eigenvectors corresponding to λ_1 and λ_2, respectively. Now normalize each chosen eigenvector to obtain an orthonormal basis $\{\mathbf{v}_1, \mathbf{v}_2\}$ for $\mathcal{R}^2$ consisting of right singular vectors of A, namely

$$\mathbf{v}_1 = \frac{1}{\sqrt{5}} \begin{bmatrix} 1 \\ 2 \end{bmatrix} \quad \text{and} \quad \mathbf{v}_2 = \frac{1}{\sqrt{5}} \begin{bmatrix} 2 \\ -1 \end{bmatrix}.$$

Third, obtain an orthonormal basis for $\mathcal{R}^3$ of left singular values of A. The first two right singular vectors can be obtained from the left singular vectors as follows:

$$\mathbf{u}_1 = \frac{1}{\sigma_1} A \mathbf{v}_1 = \frac{1}{\sqrt{7}} \begin{bmatrix} 1 & 1 \\ 1 & -1 \\ 1 & 2 \end{bmatrix} \frac{1}{\sqrt{5}} \begin{bmatrix} 1 \\ 2 \end{bmatrix} = \frac{1}{\sqrt{35}} \begin{bmatrix} 3 \\ -1 \\ 5 \end{bmatrix}$$

and

$$\mathbf{u}_2 = \frac{1}{\sigma_1} A \mathbf{v}_2 = \frac{1}{\sqrt{2}} \begin{bmatrix} 1 & 1 \\ 1 & -1 \\ 1 & 2 \end{bmatrix} \frac{1}{\sqrt{5}} \begin{bmatrix} 2 \\ -1 \end{bmatrix} = \frac{1}{\sqrt{10}} \begin{bmatrix} 1 \\ 3 \\ 0 \end{bmatrix}.$$

Fourth, choose a third left singular vector, any unit vector that is orthogonal to both $\mathbf{u}_1$ and $\mathbf{u}_2$. This can be done in two steps. Begin by finding a nonzero solution to the system of linear equations

$$\begin{aligned} 3x_1 - x_2 + 5x_3 &= 0 \\ x_1 + 3x_2 \qquad\ &= 0, \end{aligned}$$

for example, $\begin{bmatrix} -3 \\ 1 \\ 2 \end{bmatrix}$. Now divide this vector by its norm to obtain a unit vector, which

is used as the third left singular vector

$$\mathbf{u}_3 = \frac{1}{\sqrt{14}} \begin{bmatrix} -3 \\ 1 \\ 2 \end{bmatrix}.$$

Finally, let

$$U = [\mathbf{u}_1 \ \mathbf{u}_2 \ \mathbf{u}_3] = \begin{bmatrix} \frac{3}{\sqrt{35}} & \frac{1}{\sqrt{10}} & \frac{-3}{\sqrt{14}} \\ \frac{-1}{\sqrt{35}} & \frac{3}{\sqrt{10}} & \frac{1}{\sqrt{14}} \\ \frac{5}{\sqrt{35}} & 0 & \frac{2}{\sqrt{14}} \end{bmatrix}, \qquad \Sigma = \begin{bmatrix} \sigma_1 & 0 \\ 0 & \sigma_2 \\ 0 & 0 \end{bmatrix} = \begin{bmatrix} \sqrt{7} & 0 \\ 0 & \sqrt{2} \\ 0 & 0 \end{bmatrix},$$

and

$$V = [\mathbf{v}_1 \ \mathbf{v}_2] = \begin{bmatrix} \frac{1}{\sqrt{5}} & \frac{2}{\sqrt{5}} \\ \frac{2}{\sqrt{5}} & \frac{-1}{\sqrt{5}} \end{bmatrix}$$

to obtain the singular value decomposition

$$U\Sigma V^T = \begin{bmatrix} \frac{3}{\sqrt{35}} & \frac{1}{\sqrt{10}} & \frac{-3}{\sqrt{14}} \\ \frac{-1}{\sqrt{35}} & \frac{3}{\sqrt{10}} & \frac{1}{\sqrt{14}} \\ \frac{5}{\sqrt{35}} & 0 & \frac{2}{\sqrt{14}} \end{bmatrix} \begin{bmatrix} \sqrt{7} & 0 \\ 0 & \sqrt{2} \\ 0 & 0 \end{bmatrix} \begin{bmatrix} \frac{1}{\sqrt{5}} & \frac{2}{\sqrt{5}} \\ \frac{2}{\sqrt{5}} & \frac{-1}{\sqrt{5}} \end{bmatrix}^T.$$

9. First, compute

$$A^T A = \begin{bmatrix} 1 & 2 & 1 \\ 1 & 0 & -1 \\ 2 & -1 & 0 \end{bmatrix} \begin{bmatrix} 1 & 1 & 2 \\ 2 & 0 & -1 \\ 1 & -1 & 0 \end{bmatrix} = \begin{bmatrix} 6 & 0 & 0 \\ 0 & 2 & 2 \\ 0 & 2 & 5 \end{bmatrix}.$$

Second, determine the eigenvalues of $A^T A$ from the characteristic polynomial

$$\det (A^T A - I_3) = \det \begin{bmatrix} 6-t & 0 & 0 \\ 0 & 2-t & 2 \\ 0 & 2 & 5-t \end{bmatrix} = -(t-6)^2(t-1).$$

Thus $\lambda_1 = 6$, $\lambda_2 = 6$, and $\lambda_3 = 1$ are the eigenvalues (including multiplicities) of $A^T A$. The singular values of A are therefore $\sigma_1 = \sqrt{6}$, $\sigma_2 = \sqrt{6}$, and $\sigma_3 = 1$. It can be shown (we omit the details) that

$$\left\{ \begin{bmatrix} 1 \\ 0 \\ 0 \end{bmatrix}, \frac{1}{\sqrt{5}} \begin{bmatrix} 0 \\ 1 \\ 2 \end{bmatrix}, \frac{1}{\sqrt{5}} \begin{bmatrix} 0 \\ 2 \\ -1 \end{bmatrix} \right\}$$

is an orthonormal basis for $\mathcal{R}^3$ consisting of eigenvectors corresponding to the eigenvalues λ_1, λ_2, and λ_3, respectively. Let $\mathbf{v}_1$, $\mathbf{v}_2$, and $\mathbf{v}_3$ denote the vectors in this basis. These vectors are right singular vectors of A.

Third, obtain an orthonormal basis of eigenvectors of $\mathcal{R}^3$ consisting of left singular vectors:

$$\mathbf{u}_1 = \frac{1}{\sigma_1} A\mathbf{v}_1 = \frac{1}{\sqrt{6}} \begin{bmatrix} 1 & 1 & 2 \\ 2 & 0 & -1 \\ 1 & -1 & 0 \end{bmatrix} \begin{bmatrix} 1 \\ 0 \\ 0 \end{bmatrix} = \frac{1}{\sqrt{6}} \begin{bmatrix} 1 \\ 2 \\ 1 \end{bmatrix},$$

$$\mathbf{u}_2 = \frac{1}{\sigma_2} A\mathbf{v}_2 = \frac{1}{\sqrt{6}} \begin{bmatrix} 1 & 1 & 2 \\ 2 & 0 & -1 \\ 1 & -1 & 0 \end{bmatrix} \frac{1}{\sqrt{5}} \begin{bmatrix} 0 \\ 1 \\ 2 \end{bmatrix} = \frac{1}{\sqrt{30}} \begin{bmatrix} 5 \\ -2 \\ -1 \end{bmatrix}$$

and

$$\mathbf{u}_3 = \frac{1}{\sigma_3} A\mathbf{v}_3 = \frac{1}{1} \begin{bmatrix} 1 & 1 & 2 \\ 2 & 0 & -1 \\ 1 & -1 & 0 \end{bmatrix} \frac{1}{\sqrt{5}} \begin{bmatrix} 0 \\ 2 \\ -1 \end{bmatrix} = \frac{1}{\sqrt{5}} \begin{bmatrix} 0 \\ 1 \\ -2 \end{bmatrix}.$$

Finally, let

$$U = [\mathbf{u}_1 \ \mathbf{u}_2 \ \mathbf{u}_3] = \begin{bmatrix} \frac{1}{\sqrt{6}} & \frac{5}{\sqrt{30}} & 0 \\ \frac{2}{\sqrt{6}} & \frac{-2}{\sqrt{30}} & \frac{1}{\sqrt{5}} \\ \frac{1}{\sqrt{6}} & \frac{-1}{\sqrt{30}} & \frac{-2}{\sqrt{5}} \end{bmatrix}, \qquad \Sigma = \begin{bmatrix} \sigma_1 & 0 & 0 \\ 0 & \sigma_2 & 0 \\ 0 & 0 & \sigma_3 \end{bmatrix} = \begin{bmatrix} \sqrt{6} & 0 & 0 \\ 0 & \sqrt{6} & 0 \\ 0 & 0 & 1 \end{bmatrix},$$

and

$$V = [\mathbf{v}_1 \ \mathbf{v}_2 \ \mathbf{v}_3] = \begin{bmatrix} 1 & 0 & 0 \\ 0 & \frac{1}{\sqrt{5}} & \frac{2}{\sqrt{5}} \\ 0 & \frac{2}{\sqrt{5}} & \frac{-1}{\sqrt{5}} \end{bmatrix}$$

to obtain the singular value decomposition

$$\begin{bmatrix} \frac{1}{\sqrt{6}} & \frac{5}{\sqrt{30}} & 0 \\ \frac{2}{\sqrt{6}} & \frac{-2}{\sqrt{30}} & \frac{1}{\sqrt{5}} \\ \frac{1}{\sqrt{6}} & \frac{-1}{\sqrt{30}} & \frac{-2}{\sqrt{5}} \end{bmatrix} \begin{bmatrix} \sqrt{6} & 0 & 0 \\ 0 & \sqrt{6} & 0 \\ 0 & 0 & 1 \end{bmatrix} \begin{bmatrix} 1 & 0 & 0 \\ 0 & \frac{1}{\sqrt{5}} & \frac{2}{\sqrt{5}} \\ 0 & \frac{2}{\sqrt{5}} & \frac{-1}{\sqrt{5}} \end{bmatrix}^T.$$

13. From the given characteristic polynomial, we see that the eigenvalues of $A^T A$ (including multiplicities) are 21, 18, 0, and 0. Thus the (nonzero) singular values of A are $\sigma_1 = \sqrt{21}$ and $\sigma_2 = \sqrt{18}$. It can be shown (we omit the details) that

$$\left\{ \frac{1}{\sqrt{7}} \begin{bmatrix} 1 \\ 2 \\ 1 \\ 1 \end{bmatrix}, \frac{1}{\sqrt{3}} \begin{bmatrix} 1 \\ -1 \\ 0 \\ 1 \end{bmatrix}, \frac{1}{11} \begin{bmatrix} 1 \\ 1 \\ -3 \\ 0 \end{bmatrix}, \frac{1}{\sqrt{2}} \begin{bmatrix} 1 \\ 0 \\ 0 \\ -1 \end{bmatrix} \right\}$$

is an orthonormal basis for $\mathcal{R}^4$ consisting of eigenvectors of $A^T A$ corresponding to the eigenvalues 21, 18, 0, and 0, respectively. Let $\mathbf{v}_1$, $\mathbf{v}_2$, $\mathbf{v}_3$, and $\mathbf{v}_4$ denote the vectors in this basis. These vectors constitute a set of right singular vectors of A. To obtain left singular vectors, let

$$\mathbf{u}_1 = \frac{1}{\sigma_1} A\mathbf{v}_1 = \frac{1}{\sqrt{21}} \begin{bmatrix} 3 & 0 & 1 & 3 \\ 0 & 3 & 1 & 0 \\ 0 & -3 & -1 & 0 \end{bmatrix} \frac{1}{\sqrt{7}} \begin{bmatrix} 1 \\ 2 \\ 1 \\ 1 \end{bmatrix} = \frac{1}{\sqrt{3}} \begin{bmatrix} 1 \\ 1 \\ -1 \end{bmatrix}$$

and

$$\mathbf{u}_2 = \frac{1}{\sigma_2} A\mathbf{v}_2 = \frac{1}{\sqrt{18}} \begin{bmatrix} 3 & 0 & 1 & 3 \\ 0 & 3 & 1 & 0 \\ 0 & -3 & -1 & 0 \end{bmatrix} \frac{1}{\sqrt{3}} \begin{bmatrix} 1 \\ -1 \\ 0 \\ 1 \end{bmatrix} = \frac{1}{\sqrt{6}} \begin{bmatrix} 2 \\ -1 \\ 1 \end{bmatrix}.$$

Since $\mathbf{u}_1$ and $\mathbf{u}_2$ are orthonormal, the set of these vectors can be extended to an orthonormal basis $\{\mathbf{u}_1, \mathbf{u}_2, \mathbf{u}_3\}$ for $\mathcal{R}^3$. The only requirements for $\mathbf{u}_3$ are that it be a unit vector that is orthogonal to both $\mathbf{u}_1$ and $\mathbf{u}_2$. For example, choose

$$\mathbf{u}_3 = \frac{1}{\sqrt{2}} \begin{bmatrix} 0 \\ 1 \\ 1 \end{bmatrix}.$$

Finally, let

$$U = [\mathbf{u}_1 \ \mathbf{u}_2 \ \mathbf{u}_3] = \begin{bmatrix} \frac{1}{\sqrt{3}} & \frac{2}{\sqrt{6}} & 0 \\ \frac{1}{\sqrt{3}} & \frac{-1}{\sqrt{6}} & \frac{1}{\sqrt{2}} \\ \frac{-1}{\sqrt{3}} & \frac{1}{\sqrt{6}} & \frac{1}{\sqrt{2}} \end{bmatrix}, \Sigma = \begin{bmatrix} \sigma_1 & 0 & 0 & 0 \\ 0 & \sigma_2 & 0 & 0 \\ 0 & 0 & 0 & 0 \end{bmatrix} = \begin{bmatrix} \sqrt{21} & 0 & 0 & 0 \\ 0 & \sqrt{18} & 0 & 0 \\ 0 & 0 & 0 & 0 \end{bmatrix}$$

and

$$V = [\mathbf{v}_1 \ \mathbf{v}_2 \ \mathbf{v}_3 \ \mathbf{v}_4] = \begin{bmatrix} \frac{1}{\sqrt{7}} & \frac{1}{\sqrt{3}} & \frac{1}{\sqrt{11}} & \frac{1}{\sqrt{2}} \\ \frac{2}{\sqrt{7}} & \frac{-1}{\sqrt{3}} & \frac{1}{\sqrt{11}} & 0 \\ \frac{1}{\sqrt{7}} & 0 & \frac{-3}{\sqrt{11}} & 0 \\ \frac{1}{\sqrt{7}} & \frac{1}{\sqrt{3}} & 0 & \frac{-1}{\sqrt{2}} \end{bmatrix}$$

to obtain the singular value decomposition

$$\begin{bmatrix} \frac{1}{\sqrt{3}} & \frac{2}{\sqrt{6}} & 0 \\ \frac{1}{\sqrt{3}} & \frac{-1}{\sqrt{6}} & \frac{1}{\sqrt{2}} \\ \frac{-1}{\sqrt{3}} & \frac{1}{\sqrt{6}} & \frac{1}{\sqrt{2}} \end{bmatrix} \begin{bmatrix} \sqrt{21} & 0 & 0 & 0 \\ 0 & \sqrt{18} & 0 & 0 \\ 0 & 0 & 0 & 0 \end{bmatrix} \begin{bmatrix} \frac{1}{\sqrt{7}} & \frac{1}{\sqrt{3}} & \frac{1}{\sqrt{11}} & \frac{1}{\sqrt{2}} \\ \frac{2}{\sqrt{7}} & \frac{-1}{\sqrt{3}} & \frac{1}{\sqrt{11}} & 0 \\ \frac{1}{\sqrt{7}} & 0 & \frac{-3}{\sqrt{11}} & 0 \\ \frac{1}{\sqrt{7}} & \frac{1}{\sqrt{3}} & 0 & \frac{-1}{\sqrt{2}} \end{bmatrix}^T.$$

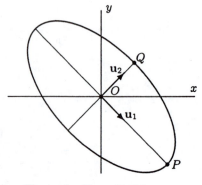

15. In the figure at the right:

$$\mathbf{u}_1 = \frac{1}{\sqrt{2}} \begin{bmatrix} 1 \\ -1 \end{bmatrix}, \ \mathbf{u}_2 = \frac{1}{\sqrt{2}} \begin{bmatrix} 1 \\ 1 \end{bmatrix},$$

$OP = 2\sqrt{2}$, and $OQ = \sqrt{2}$.

Figure for Exercise 15

17. Let $\mathbf{z}$ be the solution to this problem and $\mathbf{b} = \begin{bmatrix} 2 \\ 4 \end{bmatrix}$. If

$$A = \begin{bmatrix} 1 & 1 \\ 2 & 2 \end{bmatrix}$$

is the coefficient matrix of the system, then

$$A^T A = \begin{bmatrix} 1 & 2 \\ 1 & 2 \end{bmatrix} \begin{bmatrix} 1 & 1 \\ 2 & 2 \end{bmatrix} = \begin{bmatrix} 5 & 5 \\ 5 & 5 \end{bmatrix}.$$

Since $A^T A$ has the characteristic polynomial

$$\det (A^T A - t I_2) = \det \begin{bmatrix} 5 - t & 5 \\ 5 & 5 - t \end{bmatrix} = t(t - 10),$$

it has the eigenvalues 10 and 0. Therefore $\sigma_1 = \sqrt{10}$ is the (nonzero) singular value of A. An orthonormal basis of eigenvectors corresponding to 10 and 0, respectively, is

$$\left\{ \frac{1}{\sqrt{2}} \begin{bmatrix} 1 \\ 1 \end{bmatrix}, \frac{1}{\sqrt{2}} \begin{bmatrix} 1 \\ -1 \end{bmatrix} \right\}.$$

Let $\mathbf{v}_1$ and $\mathbf{v}_2$ denote the vectors in this set. Next, let

$$\mathbf{u}_1 = \frac{1}{\sigma_1} A \mathbf{v}_1 = \frac{1}{\sqrt{10}} \begin{bmatrix} 1 & 1 \\ 2 & 2 \end{bmatrix} \frac{1}{\sqrt{2}} \begin{bmatrix} 1 \\ 1 \end{bmatrix} = \frac{1}{\sqrt{5}} \begin{bmatrix} 1 \\ 2 \end{bmatrix}.$$

Then choose a unit vector $\mathbf{u}_2$ orthogonal to $\mathbf{u}_1$, for example,

$$\mathbf{u}_2 = \frac{1}{\sqrt{5}} \begin{bmatrix} 2 \\ -1 \end{bmatrix}.$$

Let

$$U = [\mathbf{u}_1 \ \mathbf{u}_2] = \frac{1}{\sqrt{5}} \begin{bmatrix} 1 & 2 \\ 2 & -1 \end{bmatrix}, \qquad \Sigma = \begin{bmatrix} \sigma_1 & 0 \\ 0 & 0 \end{bmatrix} = \begin{bmatrix} \sqrt{10} & 0 \\ 0 & 0 \end{bmatrix},$$

and

$$V = [\mathbf{u}_1 \ \mathbf{u}_2] = \frac{1}{\sqrt{2}} \begin{bmatrix} 1 & 1 \\ 1 & -1 \end{bmatrix}.$$

Then $A = U\Sigma V^T$ is a singular value decomposition of A. So by the boxed result on page 373,

$$\mathbf{z} = V\Sigma^\dagger U^T \mathbf{b} = \frac{1}{\sqrt{2}} \begin{bmatrix} 1 & 1 \\ 1 & -1 \end{bmatrix} \begin{bmatrix} \frac{1}{\sqrt{10}} & 0 \\ 0 & 0 \end{bmatrix} \frac{1}{\sqrt{5}} \begin{bmatrix} 1 & 2 \\ 2 & -1 \end{bmatrix}^T \begin{bmatrix} 2 \\ 4 \end{bmatrix} = \begin{bmatrix} 1 \\ 1 \end{bmatrix}.$$

21. Let $\mathbf{z}$ be the solution to this problem, and let

$$A = \begin{bmatrix} 1 & 1 \\ 1 & -1 \\ 1 & 2 \end{bmatrix} \qquad \text{and} \qquad \mathbf{b} = \begin{bmatrix} 3 \\ 1 \\ 2 \end{bmatrix}$$

be the coefficient matrix and the constant vector, respectively, for this system. By Exercise 5, $A = U\Sigma V^T$ is a singular value decomposition of A, where

$$U = \begin{bmatrix} \frac{3}{\sqrt{35}} & \frac{1}{\sqrt{10}} & \frac{-3}{\sqrt{14}} \\ \frac{-1}{\sqrt{35}} & \frac{3}{\sqrt{10}} & \frac{1}{\sqrt{14}} \\ \frac{5}{\sqrt{35}} & 0 & \frac{2}{\sqrt{14}} \end{bmatrix}, \qquad \Sigma = \begin{bmatrix} \sqrt{7} & 0 \\ 0 & \sqrt{2} \\ 0 & 0 \end{bmatrix}, \qquad \text{and} \qquad V = \begin{bmatrix} \frac{1}{\sqrt{5}} & \frac{2}{\sqrt{5}} \\ \frac{2}{\sqrt{5}} & \frac{-1}{\sqrt{5}} \end{bmatrix}.$$

Therefore, by the boxed result on page 373,

$$\mathbf{z} = V\Sigma^\dagger U^T \mathbf{b} = \begin{bmatrix} \frac{1}{\sqrt{5}} & \frac{2}{\sqrt{5}} \\ \frac{2}{\sqrt{5}} & \frac{-1}{\sqrt{5}} \end{bmatrix} \begin{bmatrix} \frac{1}{\sqrt{7}} & 0 & 0 \\ 0 & \frac{1}{\sqrt{2}} & 0 \end{bmatrix} \begin{bmatrix} \frac{3}{\sqrt{35}} & \frac{1}{\sqrt{10}} & \frac{-3}{\sqrt{14}} \\ \frac{-1}{\sqrt{35}} & \frac{3}{\sqrt{10}} & \frac{1}{\sqrt{14}} \\ \frac{5}{\sqrt{35}} & 0 & \frac{2}{\sqrt{14}} \end{bmatrix}^T \begin{bmatrix} 3 \\ 1 \\ 2 \end{bmatrix} = \frac{1}{7} \begin{bmatrix} 12 \\ 3 \end{bmatrix}.$$

25. Let $A = \begin{bmatrix} 1 \\ 2 \\ 2 \end{bmatrix}$. By Exercise 3, $A = U\Sigma V^T$ is a singular value decomposition of A, where

$$U = \begin{bmatrix} \frac{1}{3} & \frac{2}{\sqrt{5}} & \frac{2}{3\sqrt{5}} \\ \frac{2}{3} & \frac{-1}{\sqrt{5}} & \frac{4}{3\sqrt{5}} \\ \frac{2}{3} & 0 & \frac{-5}{3\sqrt{5}} \end{bmatrix}, \qquad \Sigma = \begin{bmatrix} 3 \\ 0 \\ 0 \end{bmatrix}, \qquad \text{and} \qquad V = [1].$$

Thus

$$A^\dagger = V\Sigma^\dagger U^T = [1]\begin{bmatrix} \frac{1}{3} & 0 & 0 \end{bmatrix}\begin{bmatrix} \frac{1}{3} & \frac{2}{3} & \frac{2}{3} \\ \frac{2}{\sqrt{5}} & \frac{-1}{\sqrt{5}} & 0 \\ \frac{2}{3\sqrt{5}} & \frac{4}{3\sqrt{5}} & \frac{-5}{3\sqrt{5}} \end{bmatrix} = \frac{1}{9}[1\ 2\ 2].$$

29. Let $A = \begin{bmatrix} 1 & 1 & 1 \\ 1 & -1 & -1 \end{bmatrix}$. By Exercise 7, $A = U\Sigma V^T$ is a singular value decomposition of A, where

$$U = \begin{bmatrix} \frac{1}{\sqrt{2}} & \frac{1}{\sqrt{2}} \\ \frac{-1}{\sqrt{2}} & \frac{1}{\sqrt{2}} \end{bmatrix}, \qquad \Sigma = \begin{bmatrix} 2 & 0 & 0 \\ 0 & \sqrt{2} & 0 \end{bmatrix}, \qquad \text{and} \qquad V = \begin{bmatrix} 0 & 1 & 0 \\ \frac{1}{\sqrt{2}} & 0 & \frac{1}{\sqrt{2}} \\ \frac{1}{\sqrt{2}} & 0 & \frac{-1}{\sqrt{2}} \end{bmatrix}.$$

Thus

$$A^\dagger = V\Sigma^\dagger U^T = \begin{bmatrix} 0 & 1 & 0 \\ \frac{1}{\sqrt{2}} & 0 & \frac{1}{\sqrt{2}} \\ \frac{1}{\sqrt{2}} & 0 & \frac{-1}{\sqrt{2}} \end{bmatrix}\begin{bmatrix} \frac{1}{2} & 0 \\ 0 & \frac{1}{\sqrt{2}} \\ 0 & 0 \end{bmatrix}\begin{bmatrix} \frac{1}{\sqrt{2}} & \frac{-1}{\sqrt{2}} \\ \frac{1}{\sqrt{2}} & \frac{1}{\sqrt{2}} \end{bmatrix} = \frac{1}{4}\begin{bmatrix} 2 & 2 \\ 1 & -1 \\ 1 & -1 \end{bmatrix}.$$

33. Let

$$A = \begin{bmatrix} 1 & 1 \\ 1 & -1 \\ 1 & 2 \end{bmatrix}.$$

By Exercise 5, $A = U\Sigma V^T$ is a singular value decomposition of A, where

$$U = \begin{bmatrix} \frac{3}{\sqrt{35}} & \frac{1}{\sqrt{10}} & \frac{-3}{\sqrt{14}} \\ \frac{-1}{\sqrt{35}} & \frac{3}{\sqrt{10}} & \frac{1}{\sqrt{14}} \\ \frac{5}{\sqrt{35}} & 0 & \frac{2}{\sqrt{14}} \end{bmatrix}.$$

Since rank $A = 2$, let

$$D = \begin{bmatrix} 1 & 0 & 0 \\ 0 & 1 & 0 \\ 0 & 0 & 0 \end{bmatrix}.$$

By (14), we have

$$P_W = UDU^T = \begin{bmatrix} \frac{3}{\sqrt{35}} & \frac{1}{\sqrt{10}} & \frac{-3}{\sqrt{14}} \\ \frac{-1}{\sqrt{35}} & \frac{3}{\sqrt{10}} & \frac{1}{\sqrt{14}} \\ \frac{5}{\sqrt{35}} & 0 & \frac{2}{\sqrt{14}} \end{bmatrix}\begin{bmatrix} 1 & 0 & 0 \\ 0 & 1 & 0 \\ 0 & 0 & 0 \end{bmatrix}\begin{bmatrix} \frac{3}{\sqrt{35}} & \frac{-1}{\sqrt{35}} & \frac{5}{\sqrt{35}} \\ \frac{1}{\sqrt{10}} & \frac{3}{\sqrt{10}} & 0 \\ \frac{-3}{\sqrt{14}} & \frac{1}{\sqrt{14}} & \frac{2}{\sqrt{14}} \end{bmatrix}$$

$$= \frac{1}{14} \begin{bmatrix} 5 & 3 & 6 \\ 3 & 13 & -2 \\ 6 & -2 & 10 \end{bmatrix}.$$

37. (a) Let $\mathcal{B}_1 = \{\mathbf{v}_1, \mathbf{v}_2, \ldots, \mathbf{v}_n\}$ and $\mathcal{B}_2 = \{\mathbf{u}_1, \mathbf{u}_2, \ldots, \mathbf{u}_m\}$ be the set of columns of V and U, respectively. Since V and U are orthogonal matrices, these sets are orthonormal bases for $\mathcal{R}^n$ and $\mathcal{R}^m$, respectively. Since $A = U\Sigma V^T$, we have that $AV = \Sigma U$. It follows that

$$[A\mathbf{v}_1 \; A\mathbf{v}_2 \; \cdots \; A\mathbf{v}_n] = AV$$
$$= U\Sigma$$
$$= [\sigma_1 \mathbf{u}_1 \; \sigma_2 \mathbf{u}_2 \; \cdots \; \sigma_k \mathbf{u}_k \; 0 \; \cdots \; 0],$$

and hence $A\mathbf{v}_i = \sigma_i \mathbf{u}_i$ if $i \le k$ and $A\mathbf{v}_i = 0$ if $i > k$.
To prove that $A^T \mathbf{u}_i = \sigma_i \mathbf{v}_i$ if $i \le k$ and $A^T \mathbf{u}_i = 0$ if $i > k$, apply the reasoning in (a) to

$$A^T = (U\Sigma V^T)^T = V\Sigma^T U^T.$$

It follows that $\sigma_1, \sigma_2, \ldots, \sigma_k$ are the singular values of A, and that $\mathcal{B}_1$ and $\mathcal{B}_2$ are orthonormal bases of right and left singular vectors, respectively.
(b) See (a).
(c) See (a).

41. Suppose that A is an $n \times n$ invertible matrix. For any $\mathbf{b}$ in $\mathcal{R}^n$, $A^{-1}\mathbf{b}$ is the unique solution to the system $A\mathbf{x} = \mathbf{b}$. Because $A^\dagger \mathbf{b}$ is also a solution to this system, we have $A^\dagger \mathbf{b} = A^{-1}\mathbf{b}$. Therefore $A^\dagger = A^{-1}$ by Theorem 1.2(e).

45. (a) Since Q is invertible, it has rank n, and hence Q has n singular values. By Exercise 35, the square of each singular value is an eigenvalue of $Q^T Q = I_n$. But 1 is the only eigenvalue of I_n, and hence each singular value of Q is $\sqrt{1} = 1$.
(b) $Q = QI_n I_n^T$

49. Suppose that the $m \times n$ matrix A has rank k, and let $W = \operatorname{Col} A$. We show that $AA^\dagger = P_W$. Let $\mathbf{v}$ be any vector in $\mathcal{R}^m$. Then there exist unique vectors $\mathbf{w}$ in W and $\mathbf{z}$ in $W^\perp$ such that $\mathbf{v} = \mathbf{w} + \mathbf{z}$. Since $A\mathbf{x} = \mathbf{w}$ has a solution, $A^\dagger \mathbf{w}$ is a solution to this equation by the boxed result on page 453. Therefore $AA^\dagger \mathbf{w} = \mathbf{w} = P_W \mathbf{w}$.

Since $\mathbf{z}$ is in $W^\perp$, it is orthogonal to the columns of A, and hence $A^T \mathbf{z} = 0$. Let $A = U\Sigma V^T$ be a singular value decomposition of A. Then

$$AA^\dagger = U\Sigma V^T V\Sigma^\dagger U^T = U\Sigma\Sigma^\dagger U^T = U \left[\begin{array}{c|c} I_k & O \\ \hline O & O \end{array} \right] U^T$$

by Exercise 48. Hence $AA^\dagger$ is a symmetric matrix. Thus

$$AA^\dagger \mathbf{z} = (AA^\dagger)^T \mathbf{z} = (A^\dagger)^T A^T \mathbf{z} = (A^\dagger)^T \mathbf{0} = \mathbf{0} = P_W \mathbf{z}.$$

165

Therefore

$$AA^\dagger v = AA^\dagger(w + z) = AA^\dagger w + AA^\dagger z = P_W w + P_W z = P_W(w + z) = P_W v.$$

Since v is an arbitrarily chosen vector in $\mathcal{R}^m$, it follows that $AA^\dagger = P_W$.

6.7   ROTATIONS OF $\mathcal{R}^3$ AND COMPUTER GRAPHICS

1. (a) False, consider $P = \begin{bmatrix} -1 & 0 & 0 \\ 0 & 1 & 0 \\ 0 & 0 & 1 \end{bmatrix}$.

   (b) False, let P be the matrix in (a).

   (c) False, let $Q = \begin{bmatrix} -1 & 0 & 0 \\ 0 & 0 & -1 \\ 0 & 1 & 0 \end{bmatrix}$.

   (d) False, consider I_3.
   (e) True
   (f) False, consider the matrix Q in (c).
   (g) True

5. We have

$$M = P_{90°} R_{45°} = \begin{bmatrix} 1 & 0 & 0 \\ 0 & 0 & -1 \\ 0 & 1 & 0 \end{bmatrix} \begin{bmatrix} \frac{1}{\sqrt{2}} & -\frac{1}{\sqrt{2}} & 0 \\ \frac{1}{\sqrt{2}} & \frac{1}{\sqrt{2}} & 0 \\ 0 & 0 & 1 \end{bmatrix} = \frac{1}{\sqrt{2}} \begin{bmatrix} 1 & -1 & 0 \\ 0 & 0 & -\sqrt{2} \\ 1 & 1 & 0 \end{bmatrix}.$$

9. Let $v_3 = \dfrac{1}{\sqrt{2}} \begin{bmatrix} 1 \\ 0 \\ 1 \end{bmatrix}$. We must select nonzero vectors w_1 and w_2 so that w_1, w_2, and

v_3 form an orthogonal set and w_2 lies in the direction of the rotation of w_1 by 90° in a counterclockwise direction with respect to the orientation defined by v_3. First,

choose w_1 to be any nonzero vector orthogonal to v_3, say $w_1 = \begin{bmatrix} 0 \\ 1 \\ 0 \end{bmatrix}$. Then choose

w_2 to be a nonzero vector orthogonal to w_1 and v_3. Two choices that come to mind

are $\begin{bmatrix} 1 \\ 0 \\ -1 \end{bmatrix}$ and $\begin{bmatrix} -1 \\ 0 \\ 1 \end{bmatrix}$. Which should be chosen? Since

$$\det \begin{bmatrix} 0 & 1 & 1 \\ 1 & 0 & 0 \\ 0 & -1 & 1 \end{bmatrix} < 0 \quad \text{and} \quad \det \begin{bmatrix} 0 & -1 & 1 \\ 1 & 0 & 0 \\ 0 & 1 & 1 \end{bmatrix} > 0,$$

we choose $\mathbf{w}_2 = \begin{bmatrix} -1 \\ 0 \\ 1 \end{bmatrix}$ because then the determinant of the matrix whose columns are $\mathbf{w}_1$, $\mathbf{w}_2$, and $\mathbf{v}_3$, respectively, is positive. (Once we replace $\mathbf{w}_2$ by a unit vector in the same direction, we can apply Theorem 6.20.) Now let $\mathbf{v}_1 = \mathbf{w}_1$ and

$$\mathbf{v}_2 = \frac{1}{\|\mathbf{w}_2\|}\mathbf{w}_2 = \frac{1}{\sqrt{2}} \begin{bmatrix} -1 \\ 0 \\ 1 \end{bmatrix}.$$

Let

$$V = [\mathbf{v}_1 \ \ \mathbf{v}_2 \ \ \mathbf{v}_3] = \begin{bmatrix} 0 & -\frac{1}{\sqrt{2}} & \frac{1}{\sqrt{2}} \\ 1 & 0 & 0 \\ 0 & \frac{1}{\sqrt{2}} & \frac{1}{\sqrt{2}} \end{bmatrix}.$$

Then

$$P = VR_{180°}V^T = \begin{bmatrix} 0 & -\frac{1}{\sqrt{2}} & \frac{1}{\sqrt{2}} \\ 1 & 0 & 0 \\ 0 & \frac{1}{\sqrt{2}} & \frac{1}{\sqrt{2}} \end{bmatrix} \begin{bmatrix} -1 & 0 & 0 \\ 0 & -1 & 0 \\ 0 & 0 & 1 \end{bmatrix} \begin{bmatrix} 0 & 1 & 0 \\ -\frac{1}{\sqrt{2}} & 0 & \frac{1}{\sqrt{2}} \\ \frac{1}{\sqrt{2}} & 0 & \frac{1}{\sqrt{2}} \end{bmatrix} = \begin{bmatrix} 0 & 0 & 1 \\ 0 & -1 & 0 \\ 1 & 0 & 0 \end{bmatrix}.$$

13. Let $\mathbf{v}_3 = \begin{bmatrix} 1 \\ -1 \\ 0 \end{bmatrix}$. As in Exercise 9, we select nonzero vectors $\mathbf{w}_1$ and $\mathbf{w}_2$ that are orthogonal to $\mathbf{v}_3$ and to each other so that

$$\det [\mathbf{w}_1 \ \ \mathbf{w}_2 \ \ \mathbf{v}_3] > 0.$$

We choose

$$\mathbf{w}_1 = \begin{bmatrix} 1 \\ 1 \\ 0 \end{bmatrix} \qquad \text{and} \qquad \mathbf{w}_2 = \begin{bmatrix} 0 \\ 0 \\ 1 \end{bmatrix}.$$

Next, set

$$\mathbf{v}_1 = \frac{1}{\|\mathbf{w}_1\|}\mathbf{w}_1 = \frac{1}{\sqrt{2}} \begin{bmatrix} 1 \\ 1 \\ 0 \end{bmatrix},$$

$\mathbf{v}_2 = \mathbf{w}_2$, and

$$V = [\mathbf{v}_1 \ \ \mathbf{v}_2 \ \ \mathbf{v}_3] = \begin{bmatrix} \frac{1}{\sqrt{2}} & 0 & \frac{1}{\sqrt{2}} \\ \frac{1}{\sqrt{2}} & 0 & -\frac{1}{\sqrt{2}} \\ 0 & 1 & 0 \end{bmatrix}.$$

Then

$$P = V R_{30°} V^T = \begin{bmatrix} \frac{1}{\sqrt{2}} & 0 & \frac{1}{\sqrt{2}} \\ \frac{1}{\sqrt{2}} & 0 & -\frac{1}{\sqrt{2}} \\ 0 & 1 & 0 \end{bmatrix} \begin{bmatrix} \frac{\sqrt{3}}{2} & -\frac{1}{2} & 0 \\ \frac{1}{2} & \frac{\sqrt{3}}{2} & 0 \\ 0 & 0 & 1 \end{bmatrix} \begin{bmatrix} \frac{1}{\sqrt{2}} & \frac{1}{\sqrt{2}} & 0 \\ 0 & 0 & 1 \\ \frac{1}{\sqrt{2}} & -\frac{1}{\sqrt{2}} & 0 \end{bmatrix}$$

$$= \frac{1}{4} \begin{bmatrix} \sqrt{3}+2 & \sqrt{3}-2 & -\sqrt{2} \\ \sqrt{3}-2 & \sqrt{3}+2 & -\sqrt{2} \\ \sqrt{2} & \sqrt{2} & 2\sqrt{3} \end{bmatrix}.$$

17. Let

$$M = \frac{1}{\sqrt{2}} \begin{bmatrix} 1 & -1 & 0 \\ 0 & 0 & -\sqrt{2} \\ 1 & 1 & 0 \end{bmatrix},$$

the rotation matrix in Exercise 5.

(a) The axis of rotation is spanned by an eigenvector of M corresponding to the eigenvalue 1, and hence we seek a nonzero solution to $(M - I_3)\mathbf{x} = \mathbf{0}$. The reduced row echelon form of the augmented matrix of the system of equations given in matrix form above is

$$\begin{bmatrix} 1 & 0 & -1-\sqrt{2} & 0 \\ 0 & 1 & 1 & 0 \end{bmatrix},$$

and so the general solution of the system is

$$\begin{bmatrix} x_1 \\ x_2 \\ x_3 \end{bmatrix} = x_3 \begin{bmatrix} 1+\sqrt{2} \\ -1 \\ 1 \end{bmatrix}.$$

Thus $\begin{bmatrix} 1+\sqrt{2} \\ -1 \\ 1 \end{bmatrix}$ is a vector that spans the axis of rotation.

(b) Choose any nonzero vector that is orthogonal to the vector in (a), for example,

$$\mathbf{w} = \begin{bmatrix} 0 \\ 1 \\ 1 \end{bmatrix},$$

and let α be the angle between $\mathbf{w}$ and $M\mathbf{w}$. Notice that $\|M\mathbf{w}\| = \|\mathbf{w}\|$ because

M is an orthogonal matrix. Therefore, by Exercise 54 of Section 6.1,

$$\cos \alpha = \frac{M\mathbf{w} \cdot \mathbf{w}}{\|M\mathbf{w}\|\|\mathbf{w}\|} = \frac{\dfrac{1}{\sqrt{2}}\begin{bmatrix} 1 & -1 & 0 \\ 0 & 0 & -\sqrt{2} \\ 1 & 1 & 0 \end{bmatrix}\begin{bmatrix} 0 \\ 1 \\ 1 \end{bmatrix} \cdot \begin{bmatrix} 0 \\ 1 \\ 1 \end{bmatrix}}{\left\|\begin{bmatrix} 0 \\ 1 \\ 1 \end{bmatrix}\right\|^2}$$

$$= \frac{1}{2\sqrt{2}}\begin{bmatrix} -1 \\ -\sqrt{2} \\ 1 \end{bmatrix} \cdot \begin{bmatrix} 0 \\ 1 \\ 1 \end{bmatrix} = \frac{1 - \sqrt{2}}{2\sqrt{2}}.$$

21. We have

$$\det P_\theta = \det \begin{bmatrix} 1 & 0 & 0 \\ 0 & \cos\theta & -\sin\theta \\ 0 & \sin\theta & \cos\theta \end{bmatrix} = \det \begin{bmatrix} \cos\theta & -\sin\theta \\ \sin\theta & \cos\theta \end{bmatrix} = \cos^2\theta + \sin^2\theta = 1.$$

The other determinants are computed in a similar manner.

25. By Exercise 24, T_W is a matrix transformation, and hence T_W is linear.

29. Let T be the reflection operator, and A be the standard matrix of T. Choose a nonzero vector orthogonal to $\begin{bmatrix} 1 \\ 2 \\ 3 \end{bmatrix}$ and $\begin{bmatrix} 1 \\ 0 \\ -1 \end{bmatrix}$, for example, $\begin{bmatrix} 1 \\ -2 \\ 1 \end{bmatrix}$, and let

$$\mathcal{B} = \left\{ \begin{bmatrix} 1 \\ 2 \\ 3 \end{bmatrix}, \begin{bmatrix} 1 \\ 0 \\ -1 \end{bmatrix}, \begin{bmatrix} 1 \\ -2 \\ 1 \end{bmatrix} \right\}.$$

Then $\mathcal{B}$ is a basis for $\mathcal{R}^3$, and

$$T\left(\begin{bmatrix} 1 \\ 2 \\ 3 \end{bmatrix}\right) = \begin{bmatrix} 1 \\ 2 \\ 3 \end{bmatrix}, \qquad T\left(\begin{bmatrix} 1 \\ 0 \\ -1 \end{bmatrix}\right) = \begin{bmatrix} 1 \\ 0 \\ -1 \end{bmatrix}, \qquad \text{and} \qquad T\left(\begin{bmatrix} 1 \\ -2 \\ 1 \end{bmatrix}\right) = -\begin{bmatrix} 1 \\ -2 \\ 1 \end{bmatrix}.$$

Let B be the matrix whose columns are the vectors in $\mathcal{B}$. Then

$$[T]_\mathcal{B} = B^{-1}AB = \begin{bmatrix} 1 & 0 & 0 \\ 0 & 1 & 0 \\ 0 & 0 & -1 \end{bmatrix} = D,$$

and therefore

$$A = BDB^{-1} = \begin{bmatrix} 1 & 1 & 1 \\ 2 & 0 & -2 \\ 3 & -1 & 1 \end{bmatrix}\begin{bmatrix} 1 & 0 & 0 \\ 0 & 1 & 0 \\ 0 & 0 & -1 \end{bmatrix}\begin{bmatrix} \frac{1}{6} & \frac{1}{6} & \frac{1}{6} \\ \frac{2}{3} & \frac{1}{6} & -\frac{1}{3} \\ \frac{1}{6} & -\frac{1}{3} & \frac{1}{6} \end{bmatrix} = \frac{1}{3}\begin{bmatrix} 2 & 2 & -1 \\ 2 & -1 & 2 \\ -1 & 2 & 2 \end{bmatrix}.$$

31. Let T be the reflection operator, and let A be the standard matrix of T. First, obtain a basis for W, for example,

$$\left\{ \begin{bmatrix} 1 \\ -1 \\ 0 \end{bmatrix}, \begin{bmatrix} 1 \\ 0 \\ -1 \end{bmatrix} \right\}.$$

Next, select a vector that is orthogonal to the vectors in W, for example,

$$\begin{bmatrix} 1 \\ 1 \\ 1 \end{bmatrix}.$$

Then the set

$$\mathcal{B} = \left\{ \begin{bmatrix} 1 \\ -1 \\ 0 \end{bmatrix}, \begin{bmatrix} 1 \\ 0 \\ -1 \end{bmatrix}, \begin{bmatrix} 1 \\ 1 \\ 1 \end{bmatrix} \right\}$$

consisting of these vectors is a basis for $\mathcal{R}^3$. Furthermore,

$$T\left(\begin{bmatrix} 1 \\ -1 \\ 0 \end{bmatrix} \right) = \begin{bmatrix} 1 \\ -1 \\ 0 \end{bmatrix}, \quad T\left(\begin{bmatrix} 1 \\ 0 \\ -1 \end{bmatrix} \right) = \begin{bmatrix} 1 \\ 0 \\ -1 \end{bmatrix}, \quad \text{and} \quad T\left(\begin{bmatrix} 1 \\ 1 \\ 1 \end{bmatrix} \right) = -\begin{bmatrix} 1 \\ 1 \\ 1 \end{bmatrix}.$$

Therefore

$$[T]_\mathcal{B} = \begin{bmatrix} 1 & 0 & 0 \\ 0 & 1 & 0 \\ 0 & 0 & -1 \end{bmatrix}.$$

Let P be the matrix whose columns are the vectors in $\mathcal{B}$. Then

$$A = P[T]_\mathcal{B}P^{-1} = \begin{bmatrix} 1 & 1 & 1 \\ -1 & 0 & 1 \\ 0 & -1 & 1 \end{bmatrix} \begin{bmatrix} 1 & 0 & 0 \\ 0 & 1 & 0 \\ 0 & 0 & -1 \end{bmatrix} \begin{bmatrix} 1 & 1 & 1 \\ -1 & 0 & 1 \\ 0 & -1 & 1 \end{bmatrix}^{-1}$$

$$= \frac{1}{3} \begin{bmatrix} 1 & -2 & -2 \\ -2 & 1 & -2 \\ -2 & -2 & 1 \end{bmatrix}.$$

33. Let T be the reflection operator and A be the standard matrix of T. Find a basis for the subspace of $\mathcal{R}^3$ consisting of vectors orthogonal to $\mathbf{v}$. This subspace is the solution set to the system of linear equations

$$x_2 + 2x_2 - x_3 = 0,$$

which has the parametric representation

$$\begin{bmatrix} x_1 \\ x_2 \\ x_3 \end{bmatrix} = x_2 \begin{bmatrix} -2 \\ 1 \\ 0 \end{bmatrix} + x_3 \begin{bmatrix} 1 \\ 0 \\ 1 \end{bmatrix}.$$

It follows that

$$\left\{ \begin{bmatrix} -2 \\ 1 \\ 0 \end{bmatrix}, \begin{bmatrix} 1 \\ 0 \\ 1 \end{bmatrix} \right\}$$

is a basis for this subspace. Adjoin $\mathbf{v}$ to this set to obtain a basis

$$\mathcal{B} = \left\{ \begin{bmatrix} -2 \\ 1 \\ 0 \end{bmatrix}, \begin{bmatrix} 1 \\ 0 \\ 1 \end{bmatrix}, \begin{bmatrix} 1 \\ 2 \\ -1 \end{bmatrix} \right\}$$

for $\mathcal{R}^3$, and let B be the matrix whose columns are the vectors in $\mathcal{B}$. As in Exercise 29,

$$[T]_\mathcal{B} = B^{-1}AB = \begin{bmatrix} 1 & 0 & 0 \\ 0 & 1 & 0 \\ 0 & 0 & -1 \end{bmatrix} = D,$$

and therefore

$$A = BDB^{-1} = \begin{bmatrix} -2 & 1 & 1 \\ 0 & 1 & 2 \\ 0 & 1 & -1 \end{bmatrix} \begin{bmatrix} 1 & 0 & 0 \\ 0 & 1 & 0 \\ 0 & 0 & -1 \end{bmatrix} \begin{bmatrix} -\frac{1}{3} & \frac{1}{3} & \frac{1}{3} \\ \frac{1}{6} & \frac{1}{3} & \frac{5}{6} \\ \frac{1}{6} & \frac{1}{3} & -\frac{1}{6} \end{bmatrix} = \frac{1}{3} \begin{bmatrix} 2 & -2 & 1 \\ -2 & -1 & 2 \\ 1 & 2 & 2 \end{bmatrix}.$$

37. Let

$$D = \begin{bmatrix} 1 & 0 & 0 \\ 0 & 1 & 0 \\ 0 & 0 & -1 \end{bmatrix},$$

which is the standard matrix of the reflection about the z-axis. Recall that $R_{90°}$ is the 90°-rotation matrix about the z-axis. Let

$$C = DR_{90°} = \begin{bmatrix} 1 & 0 & 0 \\ 0 & 1 & 0 \\ 0 & 0 & -1 \end{bmatrix} \begin{bmatrix} 0 & -1 & 0 \\ 1 & 0 & 0 \\ 0 & 0 & 1 \end{bmatrix} = \begin{bmatrix} 0 & -1 & 0 \\ 1 & 0 & 0 \\ 0 & 0 & -1 \end{bmatrix}.$$

Then C is not the standard matrix of a reflection operator because -1 is not an eigenvalue of C. However, C is an orthogonal matrix because it is a product of D and $R_{90°}$, which are orthogonal matrices. Finally,

$$\det C = \det (DR_{90°}) = (\det D)(\det R_{90°}) = (-1)(1) = -1.$$

41. (a) Since

$$\det \begin{bmatrix} 1 & 0 & 0 \\ 0 & -1 & 0 \\ 0 & 0 & -1 \end{bmatrix} = 1,$$

the matrix is a rotation matrix by Theorem 6.19.

(b) Observe that $\begin{bmatrix} 1 \\ 0 \\ 0 \end{bmatrix}$ is an eigenvector of the matrix corresponding to eigenvalue 1, and therefore this vector spans the axis of rotation.

45. Let M denote the given matrix.

(a) Since

$$\det M = \det \begin{bmatrix} \frac{1}{\sqrt{2}} & 0 & \frac{1}{\sqrt{2}} \\ 0 & 1 & 0 \\ \frac{1}{\sqrt{2}} & 0 & \frac{-1}{\sqrt{2}} \end{bmatrix} = -1,$$

M is not a rotation matrix by Theorem 6.19. We can establish that M is the standard matrix of a reflection operator provided that we can show that M has a 2-dimensional eigenspace corresponding to eigenvalue 1. For under this circumstance, it must have a third eigenvector corresponding to eigenvalue -1 because the determinant is the product of its eigenvalues. The reduced row echelon form of $M - I_3$ is

$$\begin{bmatrix} 1 & 0 & -(1+\sqrt{2}) \\ 0 & 0 & 0 \\ 0 & 0 & 0 \end{bmatrix},$$

and hence the eigenspace corresponding to eigenvalue 1 is 2-dimensional. Therefore M is the standard matrix of a reflection operator.

(b) The matrix equation $(M - I_2)\mathbf{x} = \mathbf{0}$ is the solution space of the system

$$x_1 - (1+\sqrt{2})x_3 = 0,$$

and hence the parametric representation of the general solution to this system is

$$\begin{bmatrix} x_1 \\ x_2 \\ x_3 \end{bmatrix} = x_2 \begin{bmatrix} 0 \\ 1 \\ 0 \end{bmatrix} + x_3 \begin{bmatrix} 1+\sqrt{2} \\ 0 \\ 1 \end{bmatrix}.$$

It follows that

$$\left\{ \begin{bmatrix} 0 \\ 1 \\ 0 \end{bmatrix}, \begin{bmatrix} 1+\sqrt{2} \\ 0 \\ 1 \end{bmatrix} \right\}$$

is a basis for the 2-dimensional subspace about which $\mathcal{R}^2$ is reflected.

49.   (rounded to 4 places after the decimal) $\text{Span} \left\{ \begin{bmatrix} .4609 \\ .1769 \\ .8696 \end{bmatrix} \right\}$, $48°$

CHAPTER 6 REVIEW

1. (a) True
   (b) True
   (c) False, the vectors must both lie in $\mathcal{R}^n$ for some n.
   (d) True
   (e) True
   (f) True
   (g) True
   (h) True
   (i) False, if W is a 1-dimensional subspace of $\mathcal{R}^3$, then $\dim W^\perp = 2$.
   (j) False, I_n is an invertible orthogonal projection matrix.
   (k) True
   (l) False, let W be the x-axis in $\mathcal{R}^2$, and let $\mathbf{v} = \begin{bmatrix} 1 \\ 2 \end{bmatrix}$. Then $\mathbf{w} = \begin{bmatrix} 1 \\ 0 \end{bmatrix}$, which is not orthogonal to $\mathbf{v}$.
   (m) False, it minimizes the sum of the squares of the *vertical distances* from the data points to the line.
   (n) False, in addition, each column must have length equal to 1.
   (o) False, consider $\begin{bmatrix} 1 & 1 \\ 1 & 2 \end{bmatrix}$, which has determinant 1 but is not an orthogonal matrix.
   (p) True
   (q) True
   (r) False, only symmetric matrices have spectral decompositions.
   (s) True

5. (a) $\|\mathbf{u}\| = \sqrt{1^2 + (-1)^2 + 2^2} = \sqrt{6}$ and $\|\mathbf{v}\| = \sqrt{2^2 + 4^2 + 1^2} = \sqrt{21}$
   (b) $d = \|\mathbf{u} - \mathbf{v}\| = \sqrt{(1-2)^2 + (-1-4)^2 + (2-1)^2} = \sqrt{27}$
   (c) $\mathbf{u} \cdot \mathbf{v} = (1)(2) + (-1)(4) + (2)(1) = 0$
   (d) $\mathbf{u}$ and $\mathbf{v}$ are orthogonal.

9. $(2\mathbf{u} + 3\mathbf{v}) \cdot \mathbf{w} = 2(\mathbf{u} \cdot \mathbf{w}) + 3(\mathbf{v} \cdot \mathbf{w}) = 2(5) + 3(-3) = 1$

13. Let A denote the matrix whose columns are the vectors in S. Since the reduced row echelon form of A is

$$\begin{bmatrix} 1 & 0 & 0 \\ 0 & 1 & 0 \\ 0 & 0 & 1 \\ 0 & 0 & 0 \end{bmatrix},$$

which has rank 3, the columns of S are linearly independent. Furthermore, S is not orthogonal.

Let $\mathbf{u}_1$, $\mathbf{u}_2$, and $\mathbf{u}_3$ denote the vectors in S, listed in the same order. Set $\mathbf{v}_1 = \mathbf{u}_1$,

$$\mathbf{v}_2 = \mathbf{u}_2 - \frac{\mathbf{u}_2 \cdot \mathbf{v}_1}{\|\mathbf{v}_1\|^2}\mathbf{v}_1 = \begin{bmatrix} 0 \\ 0 \\ 1 \\ 1 \end{bmatrix} - \frac{-1}{3}\begin{bmatrix} 1 \\ 1 \\ -1 \\ 0 \end{bmatrix} = \frac{1}{3}\begin{bmatrix} 1 \\ 1 \\ 2 \\ 3 \end{bmatrix},$$

and

$$\mathbf{v}_3 = \mathbf{u}_3 - \frac{\mathbf{u}_3 \cdot \mathbf{v}_1}{\|\mathbf{v}_1\|^2}\mathbf{v}_1 - \frac{\mathbf{u}_3 \cdot \mathbf{v}_2}{\|\mathbf{v}_2\|^2}\mathbf{v}_2 = \begin{bmatrix} 1 \\ 2 \\ 0 \\ 1 \end{bmatrix} - \frac{3}{3}\begin{bmatrix} 1 \\ 1 \\ -1 \\ 0 \end{bmatrix} - \left(\frac{2}{\frac{15}{9}}\right)\left(\frac{1}{3}\right)\begin{bmatrix} 1 \\ 1 \\ 2 \\ 3 \end{bmatrix} = \frac{1}{5}\begin{bmatrix} -2 \\ 3 \\ 1 \\ -1 \end{bmatrix}.$$

Therefore the orthogonal basis is

$$\left\{ \begin{bmatrix} 1 \\ 1 \\ -1 \\ 0 \end{bmatrix}, \frac{1}{3}\begin{bmatrix} 1 \\ 1 \\ 2 \\ 3 \end{bmatrix}, \frac{1}{5}\begin{bmatrix} -2 \\ 3 \\ 1 \\ -1 \end{bmatrix} \right\}.$$

15. A vector $\mathbf{v}$ is in $S^\perp$ if and only if $\mathbf{v}$ is orthogonal to both vectors in S, which occurs if and only if $\mathbf{v}$ is a solution to the system

$$\begin{aligned} 2x_1 + x_2 - x_3 &= 0 \\ 3x_1 + 4x_2 + 2x_3 - 2x_4 &= 0. \end{aligned}$$

The reduced row echelon form of the augmented matrix of this system is

$$\begin{bmatrix} 1 & 0 & -\frac{6}{5} & \frac{2}{5} & 0 \\ 0 & 1 & \frac{7}{5} & -\frac{4}{5} & 0 \end{bmatrix},$$

from which we obtain the parametric representation of the general solution

$$\begin{bmatrix} x_1 \\ x_2 \\ x_3 \\ x_4 \end{bmatrix} = x_3 \begin{bmatrix} \frac{6}{5} \\ -\frac{7}{5} \\ 1 \\ 0 \end{bmatrix} + x_4 \begin{bmatrix} -\frac{2}{5} \\ \frac{4}{5} \\ 0 \\ 1 \end{bmatrix}.$$

The two vectors in this representation constitute a basis for $S^\perp$. However, we can multiply each of these vectors by 5 to eliminate fractions to obtain the alternate basis

$$\left\{ \begin{bmatrix} 6 \\ -7 \\ 5 \\ 0 \end{bmatrix}, \begin{bmatrix} -2 \\ 4 \\ 0 \\ 5 \end{bmatrix} \right\}.$$

17. We have

$$\mathbf{w} = (\mathbf{v}\cdot\mathbf{v}_1)\mathbf{v}_1 + (\mathbf{v}\cdot\mathbf{v}_2)\mathbf{v}_2$$

$$= \frac{5}{\sqrt{5}}\frac{1}{\sqrt{5}}\begin{bmatrix}1\\2\\0\end{bmatrix} + \frac{-9}{\sqrt{14}}\frac{1}{\sqrt{14}}\begin{bmatrix}-2\\1\\3\end{bmatrix}$$

$$= \frac{1}{14}\begin{bmatrix}32\\19\\-27\end{bmatrix},$$

and

$$\mathbf{z} = \mathbf{v} - \mathbf{w} = \begin{bmatrix}1\\2\\-3\end{bmatrix} - \frac{1}{14}\begin{bmatrix}32\\19\\-27\end{bmatrix} = \frac{1}{14}\begin{bmatrix}-18\\9\\-15\end{bmatrix}.$$

21. As in Exercise 15, we obtain a basis for W, the orthogonal complement of the given set:

$$\left\{\begin{bmatrix}-1\\-1\\1\\0\end{bmatrix}, \begin{bmatrix}0\\0\\0\\1\end{bmatrix}\right\}.$$

Since these vectors are already orthogonal, we need only replace each vector by a unit vector in the same direction to obtain an orthonormal basis

$$\{\mathbf{v}_1, \mathbf{v}_2\} = \left\{\frac{1}{\sqrt{3}}\begin{bmatrix}-1\\-1\\1\\0\end{bmatrix}, \begin{bmatrix}0\\0\\0\\1\end{bmatrix}\right\}$$

for W. Then

$$P_W = (\mathbf{v}\cdot\mathbf{v}_1)\mathbf{v}_1 + (\mathbf{v}\cdot\mathbf{v}_2)\mathbf{v}_2$$

for any vector $\mathbf{v}$ in $\mathcal{R}^4$. Therefore

$$P_W\mathbf{e}_1 = (\mathbf{e}_1\cdot\mathbf{v}_1)\mathbf{v}_1 + (\mathbf{e}_2\cdot\mathbf{v}_2)\mathbf{v}_2 = \frac{-1}{\sqrt{3}}\frac{1}{\sqrt{3}}\begin{bmatrix}-1\\-1\\-1\\0\end{bmatrix} = \frac{1}{\sqrt{3}}\begin{bmatrix}1\\1\\-1\\0\end{bmatrix}.$$

Similarly,

$$P_W\mathbf{e}_2 = \begin{bmatrix}1\\1\\-1\\0\end{bmatrix}, \qquad P_W\mathbf{e}_3 = \begin{bmatrix}-1\\-1\\1\\0\end{bmatrix}, \qquad \text{and} \qquad P_W\mathbf{e}_4 = \begin{bmatrix}0\\0\\0\\1\end{bmatrix}.$$

Therefore

$$P_W = [P_W\mathbf{e}_1 \ \ P_W\mathbf{e}_2 \ \ P_W\mathbf{e}_3 \ \ P_W\mathbf{e}_4] = \frac{1}{3}\begin{bmatrix} 1 & 1 & -1 & 0 \\ 1 & 1 & -1 & 0 \\ -1 & -1 & 1 & 0 \\ 0 & 0 & 0 & 3 \end{bmatrix}.$$

Finally,

$$\mathbf{w} = P_W\mathbf{v} = \frac{1}{3}\begin{bmatrix} 1 & 1 & -1 & 0 \\ 1 & 1 & -1 & 0 \\ -1 & -1 & 1 & 0 \\ 0 & 0 & 0 & 3 \end{bmatrix}\begin{bmatrix} 2 \\ -1 \\ 1 \\ 2 \end{bmatrix} = \begin{bmatrix} 0 \\ 0 \\ 0 \\ 2 \end{bmatrix}.$$

23. Let

$$C = \begin{bmatrix} 1 & 1 \\ 1 & 2 \\ 1 & 3 \\ 1 & 4 \\ 1 & 5 \end{bmatrix} \qquad \text{and} \qquad \mathbf{y} = \begin{bmatrix} 3.2 \\ 5.1 \\ 7.1 \\ 9.2 \\ 11.4 \end{bmatrix}.$$

Then

$$C^T C = \begin{bmatrix} 1 & 1 & 1 & 1 & 1 \\ 1 & 2 & 3 & 4 & 5 \end{bmatrix}\begin{bmatrix} 1 & 1 \\ 1 & 2 \\ 1 & 3 \\ 1 & 4 \\ 1 & 5 \end{bmatrix} = \begin{bmatrix} 5 & 15 \\ 15 & 55 \end{bmatrix},$$

and hence

$$\begin{bmatrix} c \\ v \end{bmatrix} = (C^T C)^{-1} C^T \mathbf{y}$$

$$= \begin{bmatrix} 5 & 15 \\ 15 & 55 \end{bmatrix}^{-1}\begin{bmatrix} 1 & 1 & 1 & 1 & 1 \\ 1 & 2 & 3 & 4 & 5 \end{bmatrix}\begin{bmatrix} 3.2 \\ 5.1 \\ 7.1 \\ 9.2 \\ 11.4 \end{bmatrix}$$

$$= \frac{1}{10}\begin{bmatrix} 11 & -3 \\ -3 & 1 \end{bmatrix}\begin{bmatrix} 36.0 \\ 128.5 \end{bmatrix}$$

$$= \begin{bmatrix} 1.05 \\ 2.05 \end{bmatrix}.$$

Therefore $v = 2.05$ and $c = 1.05$.

25. $\begin{bmatrix} 0.7 & 0.3 \\ -0.3 & 0.7 \end{bmatrix} \begin{bmatrix} 0.7 & 0.3 \\ -0.3 & 0.7 \end{bmatrix}^T = \begin{bmatrix} 0.7 & 0.3 \\ -0.3 & 0.7 \end{bmatrix} \begin{bmatrix} 0.7 & -0.3 \\ 0.3 & 0.7 \end{bmatrix} = \begin{bmatrix} 0.58 & 0.00 \\ 0.00 & 0.58 \end{bmatrix} \neq \begin{bmatrix} 1 & 0 \\ 0 & 1 \end{bmatrix}$

It follows that the matrix is not orthogonal.

29. Because $\det \dfrac{1}{2} \begin{bmatrix} 1 & \sqrt{3} \\ -\sqrt{3} & 1 \end{bmatrix} = \frac{1}{4}(1+3) = 1$, the matrix is a rotation matrix. **Comparing** the first column of this matrix with the first column of the rotation matrix A_θ, we see that

$$A_\theta = \begin{bmatrix} \cos\theta & -\sin\theta \\ \sin\theta & \cos\theta \end{bmatrix} = \begin{bmatrix} \frac{1}{2} & \frac{\sqrt{3}}{2} \\ -\frac{\sqrt{3}}{2} & \frac{1}{2} \end{bmatrix},$$

and hence $\cos\theta = \frac{1}{2}$ and $\sin\theta = -\frac{\sqrt{3}}{2}$. Therefore $\theta = -60°$.

31. Because $\det \dfrac{1}{2} \begin{bmatrix} 1 & \sqrt{3} \\ \sqrt{3} & -1 \end{bmatrix} = \frac{1}{4}(-1-3) = -1$, the matrix is the standard **matrix** of a reflection. To find the line of reflection, we find an eigenvector of the **matrix** corresponding to eigenvalue 1 by solving the equation

$$\begin{bmatrix} \frac{1}{2}-1 & \frac{\sqrt{3}}{2} \\ \frac{\sqrt{3}}{2} & -\frac{1}{2}-1 \end{bmatrix} \begin{bmatrix} x_1 \\ x_2 \end{bmatrix} = \begin{bmatrix} 0 \\ 0 \end{bmatrix}.$$

The parametric representation of the general solution to this equation is

$$\begin{bmatrix} x_1 \\ x_2 \end{bmatrix} = x_2 \begin{bmatrix} \sqrt{3} \\ 1 \end{bmatrix},$$

and hence the equation of the line is $y = \frac{1}{\sqrt{3}}x$.

33. The standard matrix of T is given by

$$Q = \begin{bmatrix} 0 & -1 & 0 \\ 0 & 0 & 1 \\ 1 & 0 & 0 \end{bmatrix},$$

and so

$$QQ^T = \begin{bmatrix} 0 & -1 & 0 \\ 0 & 0 & 1 \\ 1 & 0 & 0 \end{bmatrix} \begin{bmatrix} 0 & 0 & 1 \\ -1 & 0 & 0 \\ 0 & 1 & 0 \end{bmatrix} = \begin{bmatrix} 1 & 0 & 0 \\ 0 & 1 & 0 \\ 0 & 0 & 1 \end{bmatrix}.$$

Hence Q is an orthogonal matrix, and T is an orthogonal operator.

35. The characteristic polynomial of A is

$$\det(A - tI_2) = \det \begin{bmatrix} 2-t & 3 \\ 3 & 2-t \end{bmatrix} = (t+1)(t-5),$$

and hence A has the eigenvalues $\lambda_1 = -1$ and $\lambda_2 = 5$. The vectors $\begin{bmatrix} -1 \\ 1 \end{bmatrix}$ and $\begin{bmatrix} 1 \\ 1 \end{bmatrix}$ are corresponding eigenvectors. Normalizing these vectors, we obtain the orthonormal basis

$$\{\mathbf{u}_1, \mathbf{u}_2\} = \left\{ \frac{1}{\sqrt{2}} \begin{bmatrix} -1 \\ 1 \end{bmatrix}, \frac{1}{\sqrt{2}} \begin{bmatrix} 1 \\ 1 \end{bmatrix} \right\}$$

consisting of eigenvectors of A. We use these eigenvectors and eigenvalues to obtain the spectral decomposition

$$A = \lambda_1 \mathbf{u}_1 \mathbf{u}_1^T + \lambda_2 \mathbf{u}_2 \mathbf{u}_2^T$$

$$= (-1)\frac{1}{\sqrt{2}} \begin{bmatrix} -1 \\ 1 \end{bmatrix} \frac{1}{\sqrt{2}}[-1 \ \ 1] + 5\frac{1}{\sqrt{2}} \begin{bmatrix} 1 \\ 1 \end{bmatrix} \frac{1}{\sqrt{2}}[1 \ \ 1]$$

$$= (-1) \begin{bmatrix} 0.5 & -0.5 \\ -0.5 & 0.5 \end{bmatrix} + 5 \begin{bmatrix} 0.5 & 0.5 \\ 0.5 & 0.5 \end{bmatrix}.$$

37. Let $a_{11} = 1$, $a_{22} = 1$, and $a_{12} = a_{21} = 3$. Thus

$$A = \begin{bmatrix} 1 & 3 \\ 3 & 1 \end{bmatrix}.$$

Now A has eigenvalues 4 and -2, and

$$\left\{ \frac{1}{\sqrt{2}} \begin{bmatrix} 1 \\ 1 \end{bmatrix}, \frac{1}{\sqrt{2}} \begin{bmatrix} -1 \\ 1 \end{bmatrix} \right\}$$

is an orthonormal basis of $\mathcal{R}^2$ consisting of corresponding eigenvectors. Choose the x'-axis to be in the direction of the first vector in this basis. Clearly this vector lies on the line $y = x$, pointing toward the first quadrant, and hence the x'-axis is the result of rotating the x-axis by $45°$.

The equation of this conic in $x'y'$-coordinates is

$$\lambda_1(x')^2 + \lambda_2(y')^2 = 4(x')^2 - 2(y')^2 = 16,$$

or

$$\frac{(x')^2}{4} - \frac{(y')^2}{8} = 1.$$

The conic is a hyperbola.

Chapter 7

Vector Spaces

7.1   VECTOR SPACES AND THEIR SUBSPACES

1. (a) True
   (b) False, by Theorem 7.2, the zero vector of a vector space is unique.
   (c) False if $a = 0$ and $\mathbf{v} \neq \mathbf{0}$.
   (d) True
   (e) False, any two polynomials can be added.
   (f) False. For example, if $n = 1$, $p(x) = 1 + x$, and $q(x) = 1 - x$, then $p(x)$ and $q(x)$ each have degree 1, but $p(x) + q(x)$ has degree 0.
   (g) True
   (h) True
   (i) True
   (j) True
   (k) True
   (l) False, the empty set contains no zero vector.
   (m) True

3. We verify a few of the axioms of a vector space.

   Axiom 1 Let f and g be in V. Then for any s in S,

   $$(f + g)(s) = f(s) + g(s) = g(s) + f(s) = (g + f)(s),$$

   and hence $f + g = g + f$.

   Axiom 3 Let O be the function defined by $O(s) = 0$ for all s in S. Then for any f in V and s in S, we have

   $$(f + O)(s) = f(s) + O(s) = f(s) + \mathbf{0} = f(s),$$

   and hence $f + O = f$.

   Axiom 8 Let f be in V and a and b be scalars. Then for any s in S, we have

   $$[(a + b)f](s) = (a + b) \cdot f(s) = af(s) + bf(s) = (af)(s) + (bf)(s) = (af + bf)(s),$$

   and hence $(a + b)f = af + bf$.

5. In view of Theorem 1.1, it suffices to show that the sum of two matrices in V is also in V; the product of a scalar and a matrix in V is also in V; the 2×2 zero matrix is in V; and if A is in V, then $-A$ is in V.

Suppose that A_1 and A_2 are in V, where

$$A_1 = \begin{bmatrix} a_1 & 2a_1 \\ b_1 & -b_1 \end{bmatrix} \quad \text{and} \quad A_2 = \begin{bmatrix} a_2 & 2a_2 \\ b_2 & -b_2 \end{bmatrix}$$

for scalars a_1, a_2, b_1, and b_2. Then

$$A_1 + A_2 = \begin{bmatrix} a_1 + a_2 & 2(a_1 + a_2) \\ b_1 + b_2 & -(b_1 + b_2) \end{bmatrix},$$

which is in V. For any scalar c,

$$cA_1 = \begin{bmatrix} ca_1 & 2ca_1 \\ cb_1 & -cb_1 \end{bmatrix},$$

which is in V. Setting $a = b = 0$, we see that O is in V. For any A in V, $-A = (-1)A$, and hence $-A$ is in V.

9. We have

$$(1 + 2) \odot (0, 3) = 3 \odot (0, 3) = (0, 27)$$

and

$$(1 \odot (0, 3)) \oplus (2 \odot (0, 3)) = (0, 3) \oplus (0, 12) = (0, 15).$$

So axiom 8 fails, and V is not a vector space.

13. Let f be in $\mathcal{F}(S)$, and let $-f$ be the function in $\mathcal{F}(S)$ defined by $(-f)(s) = -f(s)$ for all s in S. Then for any s in S, we have

$$(f + (-f))(s) = f(s) + (-f)(s) = f(s) - f(s) = 0 = \mathbf{0}(s),$$

and hence $f + (-f) = \mathbf{0}$.

17. We verify a few of the axioms of a vector space, and leave the others to the reader.

Axiom 1 Let T and U be in $\mathcal{L}(\mathcal{R}^n, \mathcal{R}^m)$. Then for any $\mathbf{v}$ in $\mathcal{R}^n$, we have

$$(T + U)(\mathbf{v}) = T(\mathbf{v}) + U(\mathbf{v}) = U(\mathbf{v}) + T(\mathbf{v}) = (U + T)(\mathbf{v}),$$

and hence $T + U = U + T$.

Axiom 3 Let T be in $\mathcal{L}(\mathcal{R}^n)$. Then for any $\mathbf{v}$ in $\mathcal{R}^n$, we have

$$(T + T_0)(\mathbf{v}) = T(\mathbf{v}) + T_0(\mathbf{v}) = T(\mathbf{v}) + \mathbf{0} = T(\mathbf{v}),$$

and hence $T + T_0 = T$.

Axiom 7 Let T and U be in $\mathcal{L}(\mathcal{R}^n, \mathcal{R}^m)$, and let a be a scalar. Then for any $\mathbf{v}$ in $\mathcal{R}^n$, we have

$$
\begin{aligned}
(a(T + U))(\mathbf{v}) &= a((T + U)(\mathbf{v})) \\
&= a(T(\mathbf{v}) + U(\mathbf{v})) \\
&= a(T(\mathbf{v})) + a(U(\mathbf{v})) \\
&= (aT)(\mathbf{v}) + (aU)(\mathbf{v}) \\
&= (aT + aU)(\mathbf{v}),
\end{aligned}
$$

and hence $a(T + U) = aT + aU$.

21. We apply Theorem 7.2(g) and axiom 6 to obtain

$$
(-a)\mathbf{v} = ((-1)a)\mathbf{v} = (-1)(a\mathbf{v}) = -(a\mathbf{v}),
$$

and

$$
(-a)\mathbf{v} = (a(-1))\mathbf{v} = a((-1)\mathbf{v}) = a(-\mathbf{v}).
$$

25. $(-c)(-\mathbf{v}) = (-c)((-1)\mathbf{v}) = ((-c)(-1))\mathbf{v} = c\mathbf{v}$

27. No, V is not closed under addition. For example, let $n = 2$, $A = \begin{bmatrix} 1 & 0 \\ 0 & 0 \end{bmatrix}$, and $B = \begin{bmatrix} 0 & 0 \\ 0 & 1 \end{bmatrix}$. Then $\det A = \det B = 0$, and hence A and B are in V, but $\det (A + B) = 1$, and so $A + B$ is not in V.

29. Yes. Since $OB = BO = O$, the zero matrix is in V. Now suppose that A and C are in V. Then $(A + C)B = AB + CB = BA + BC = B(A + C)$, and hence $A + C$ is in V. So V is closed under addition. Now let c be any scalar. Then $(cA)B = c(AB) = c(BA) = B(cA)$, and hence cA is in V. Therefore V is closed under scalar multiplication.

33. No, V is not closed under addition. Let $m = 4$, $p(x) = 1 + x^2 + x^4$ and $q(x) = 1 - x^2 + x^4$. Both $p(x)$ and $q(x)$ are in V, but $p(x) + q(x) = 2 + 0x^2 + 2x^4$ is not in V.

35. Yes. Since the zero polynomial is in V, it suffices to show that V is closed under addition and scalar multiplication. Suppose $p(x) = a_0 + a_1 x + \cdots + a_m x^m$ and $q(x) = b_0 + b_1 x + \cdots + b_m x^m$ are in V. Then $a_0 + a_1 = b_0 + b_1 = 0$. Also

$$
p(x) + q(x) = (a_0 + b_0) + (a_1 + b_1)x + \cdots + (a_m + b_m)x^m,
$$

and hence

$$
(a_0 + b_0) + (a_1 + b_1) = (a_0 + a_1) + (b_0 + b_1) = 0 + 0 = 0.
$$

Therefore $p(x) + q(x)$ is in V, and V is closed under addition. Furthermore, for any scalar c,

$$cp(x) = ca_0 + (ca_1)x + \cdots + (ca_m)x^m,$$

and hence $ca_0 + ca_1 = c(a_0 + a_1) = c \cdot 0 = 0$. So $cp(x)$ is in V, and it follows that V is closed under scalar multiplication.

37. Yes. Let $\mathbf{0}$ denote the zero function. Then

$$\mathbf{0}(s_1) + \mathbf{0}(s_2) + \cdots + \mathbf{0}(s_n) = 0 + 0 + \cdots + 0 = 0,$$

and hence the zero function is in V. Suppose that f and g are in V. Then

$$\begin{aligned}
(f + g)(s_1) + \cdots + (f + g)(s_n) &= (f(s_1) + g(s_1)) + \cdots + (f(s_n) + g(s_n)) \\
&= f(s_1) + f(s_2) + \cdots + f(s_n) \\
&\quad + g(s_1) + g(s_2) + \cdots + g(s_n) \\
&= 0 + 0 \\
&= 0,
\end{aligned}$$

and hence $f + g$ is in V. Thus V is closed under addition. Let c be any scalar. Then

$$\begin{aligned}
(cf)(s_1) + (cf)(s_2) + \cdots + (cf)(s_n) &= cf(s_1) + cf(s_2) + \cdots + cf(s_n) \\
&= c(f(s_1) + f(s_2) + \cdots + f(s_n)) \\
&= c(0 + 0 + \cdots + 0) \\
&= 0,
\end{aligned}$$

and hence cf is in V. Thus V is closed under scalar multiplication.

41. Since $\mathbf{0}' = \mathbf{0}$, the zero function is in S. Suppose that f and g are in S. Then $(f + g)' = f' + g' = f + g$, and hence $f + g$ is in S. Thus S is closed under addition. Let c be any scalar. Then $(cf)' = cf' = cf$, and hence cf is in S. It follows that S is closed under scalar multiplication, and therefore S is a subspace.

45. Since $\mathbf{0}$ is in both W_1 and W_2 and $\mathbf{0} = \mathbf{0} + \mathbf{0}$, it follows that $\mathbf{0}$ is in W. Now suppose that $\mathbf{u}$ and $\mathbf{v}$ are in W. Then $\mathbf{u} = \mathbf{u}_1 + \mathbf{u}_2$ and $\mathbf{v} = \mathbf{v}_1 + \mathbf{v}_2$ for some vectors $\mathbf{u}_1$ and $\mathbf{v}_1$ in W_1 and $\mathbf{u}_2$ and $\mathbf{v}_2$ in W_2. Hence

$$\mathbf{u} + \mathbf{v} = (\mathbf{u}_1 + \mathbf{u}_2) + (\mathbf{v}_1 + \mathbf{v}_2) = (\mathbf{u}_1 + \mathbf{v}_1) + (\mathbf{u}_2 + \mathbf{v}_2),$$

which lies in W because $\mathbf{u}_1 + \mathbf{v}_1$ lies in W_1 and $\mathbf{u}_2 + \mathbf{v}_2$ lies in W_2. Thus W is closed under addition. Furthermore, for any scalar a,

$$a\mathbf{u} = a(\mathbf{u}_1 + \mathbf{u}_2) = a\mathbf{u}_1 + a\mathbf{u}_2,$$

which lies in W because $a\mathbf{u}_1$ lies in W_1 and $a\mathbf{u}_2$ lies in W_2. Thus W is closed under scalar multiplication. Therefore W is a subspace of V.

7.2 DIMENSION AND ISOMORPHISM

1. **(a)** True
   (b) True
   (c) False, see Example 7.
   (d) False, no infinite-dimensional space has a finite basis.
   (e) False, it has dimension $n + 1$.
   (f) False, if V is an infinite-dimensional vector space and $\mathbf{v}$ is a nonzero vector in V, then Span $\{\mathbf{v}\}$ is a 1-dimensional subspace of V.
   (g) False, if a vector space has a finite basis of n vectors, then any subset containing more than n vectors is linearly dependent and cannot be a basis.
   (h) True
   (i) True

3. Yes. $\begin{bmatrix} 1 & 2 & 1 \\ 1 & 1 & 1 \end{bmatrix} = 1 \begin{bmatrix} 1 & 2 & 1 \\ 0 & 0 & 0 \end{bmatrix} + 1 \begin{bmatrix} 0 & 0 & 0 \\ 1 & 1 & 1 \end{bmatrix} + 0 \begin{bmatrix} 1 & 0 & 1 \\ 1 & 2 & 3 \end{bmatrix}.$

5. No. The given matrix is a linear combination if and only if the matrix equation

$$\begin{bmatrix} 2 & 2 & 2 \\ 1 & 1 & 1 \end{bmatrix} = x_1 \begin{bmatrix} 1 & 2 & 1 \\ 0 & 0 & 0 \end{bmatrix} + x_2 \begin{bmatrix} 0 & 0 & 0 \\ 1 & 1 & 1 \end{bmatrix} + x_3 \begin{bmatrix} 1 & 0 & 1 \\ 1 & 2 & 3 \end{bmatrix}$$

has a solution. Comparing the $(1,1)$-, $(1,2)$-, $(2,1)$- and $(2,2)$-entries of the right and left sides of the matrix equation yields the system of linear equations

$$\begin{aligned} x_1 \quad\ + x_3 &= 2 \\ 2x_1 \qquad\quad &= 2 \\ x_2 + x_3 &= 1 \\ x_2 + 2x_3 &= 1. \end{aligned}$$

The reduced row echelon form of the augmented matrix of this system is

$$\begin{bmatrix} 1 & 0 & 0 & 0 \\ 0 & 1 & 0 & 0 \\ 0 & 0 & 1 & 0 \\ 0 & 0 & 0 & 1 \end{bmatrix},$$

indicating that the system is inconsistent, and hence the matrix equation has no solution.

7. We show that there are no scalars a and b such that

$$a(x^3 - 2x^2 - 5x - 3) + b(3x^3 - 5x^2 - 4x - 9) = f(x).$$

Comparing the corresponding coefficients of both sides of the preceding polynomial equation, we obtain the system of linear equations

$$\begin{aligned} a + 3b &= \quad 3 \\ -2a - 5b &= -2 \\ -5a - 4b &= \quad 7 \\ -3a - 9b &= \quad 8, \end{aligned}$$

183

which has no solution.

9. Consider the matrix equation

$$x_1 \begin{bmatrix} 1 & 2 \\ 3 & 1 \end{bmatrix} + x_2 \begin{bmatrix} 1 & -5 \\ -4 & 0 \end{bmatrix} + x_3 \begin{bmatrix} 3 & -1 \\ 2 & 2 \end{bmatrix} = \begin{bmatrix} 0 & 0 \\ 0 & 0 \end{bmatrix}.$$

Comparing corresponding entries of both sides of the preceding matrix equation, we obtain the system of linear equations

$$\begin{aligned} x_1 + \ x_2 + 3x_3 &= 0 \\ 2x_1 - 5x_2 - \ x_3 &= 0 \\ 3x_1 - 4x_2 + 2x_3 &= 0 \\ x_1 \qquad\quad + 2x_3 &= 0. \end{aligned}$$

The reduced row echelon form of the augmented matrix of this equation is

$$\begin{bmatrix} 1 & 0 & 2 & 0 \\ 0 & 1 & 1 & 0 \\ 0 & 0 & 0 & 0 \\ 0 & 0 & 0 & 0 \end{bmatrix}.$$

It follows that the system has nonzero solutions. For example, setting $x_3 = 1$, we have $x_1 = -2$, $x_2 = -1$. Therefore the set of matrices is linearly dependent.

11. Notice that $\begin{bmatrix} 12 & 9 \\ -3 & 0 \end{bmatrix} = 3 \begin{bmatrix} 4 & 3 \\ -1 & 0 \end{bmatrix}$, and hence the set is linearly dependent.

13. Observe that

$$1(1+x) + 1(1-x) + (-1)(1+x^2) + (-1)(1-x^2) = 0,$$

and hence this set is linearly dependent.

15. Suppose that

$$a_1(x^2 - 2x + 5) + a_2(2x^2 - 5x + 10) + a_3 x^2 = 0.$$

Collecting like terms, we obtain

$$(a_1 + 2a_2 + a_3)x^2 + (-2a_1 - 5a_2)x + (5a_1 + 10a_2) = 0.$$

Then each coefficient is zero, and hence we obtain the system

$$\begin{aligned} a_1 + \ 2a_2 + a_3 &= 0 \\ -2a_1 - \ 5a_2 \qquad &= 0 \\ 5a_1 + 10a_2 \qquad &= 0. \end{aligned}$$

Since the only solution to this system is the zero solution, $a_1 = a_2 = a_3 = 0$, the set is linearly independent.

17. Suppose that $\{t, t\sin t\}$ is linearly dependent. Since these are both nonzero functions, there exists a nonzero scalar a such that $t\sin t = at$ for all t in $\mathcal{R}$. Setting $t = \frac{\pi}{2}$, we obtain $\frac{\pi}{2}\sin\frac{\pi}{2} = a\frac{\pi}{2}$, from which we see that $a = 1$. Setting $t = \frac{\pi}{4}$, we obtain $\frac{\pi}{4}\sin\frac{\pi}{4} = a\frac{\pi}{4}$, from which we see that $a = \frac{1}{\sqrt{2}}$. This is a contradiction, and it follows that the set is linearly independent.

19. Observe that

$$0\sin t + 1\sin^2 t + 1\cos^2 t + (-1)1 = 0$$

for all t, and hence the set is linearly dependent.

21. We show that for any positive integer n, any subset of the given set consisting of n functions is linearly independent. This is certainly true for $n = 1$ because any set consisting of a single nonzero element is linearly independent.

Now suppose that we have established that any subset consisting of k functions is linearly independent, where k is a fixed positive integer. Consider any subset consisting of $k + 1$ functions

$$\{e^{n_1 t}, e^{n_2 t}, \ldots, e^{n_{k+1} t}\}.$$

Let $a_1, a_2, \ldots, a_{k+1}$ be scalars such that

$$a_1 e^{n_1 t} + a_2 e^{n_2 t} + \cdots + a_k e^{n_k t} + a_{k+1} e^{n_{k+1} t} = 0$$

for all t. We form two equations from the equation above. The first equation is obtained by taking the derivative of both sides with respect to t, and the second equation is obtained by multiplying both sides of the equation by n_{k+1}. The results are

$$n_1 a_1 e^{n_1 t} + n_2 a_2 e^{n_2 t} + \cdots + n_k a_k e^{n_k t} + n_{k+1} a_{k+1} e^{n_{k+1} t} = 0$$

and

$$n_{k+1} a_1 e^{n_1 t} + n_{k+1} a_2 e^{n_2 t} + \cdots + n_{k+1} a_k e^{n_k t} + n_{k+1} a_{k+1} e^{n_{k+1} t} = 0.$$

Now subtract the second equation from the first equation to obtain

$$(n_1 - n_{k+1})a_1 e^{n_1 t} + (n_2 - n_{k+1})a_k e^{n_k t} + \cdots + (n_k - n_{k+1})a_k e^{n_k t} = 0.$$

Since this last equation involves a linear combination of a set of k functions, which is assumed to be linearly independent, each coefficient $(n_i - n_{k+1})a_i$ is zero. But $n_i \neq n_{k+1}$ for $i = 1, 2, \ldots, k$ and hence $a_i = 0$ for $i = 1, 2, \ldots, k$. Thus the original equation reduces to $a_{k+1} e^{n_{k+1} t} = 0$, from which we conclude that $a_{k+1} = 0$. It follows that any subset consisting of $k + 1$ functions is linearly independent.

Now apply this result $n - 1$ times to show that any subset of n functions is linearly independent. That is, since a subset of 1 function is linearly independent, the result implies that a subset of 2 functions is linearly independent. By the result, we obtain that any subset of 3 functions is linearly independent. Continuing, we eventually conclude that any subset of n functions is linearly independent.

25. Let

$$p_1(x) = \frac{(x-1)(x-2)}{(0-1)(1-2)}, \quad p_2(x) = \frac{(x-0)(x-2)}{(1-0)(1-2)}, \quad \text{and} \quad p_3(x) = \frac{(x-0)(x-1)}{(2-0)(2-1)}.$$

Then

$$p(x) = 1p_1(x) + 0p_2(x) + 3p_3(x)$$
$$= \frac{1}{2}(x-1)(x-2) + \frac{3}{2}x(x-1) = 2x^2 - 3x + 1.$$

29. A matrix $A = \begin{bmatrix} x_1 & x_2 \\ x_3 & x_4 \end{bmatrix}$ is in W if and only if $x_1 + x_4 = 0$. So for any such matrix A,

$$A = \begin{bmatrix} x_1 & x_2 \\ x_3 & -x_1 \end{bmatrix} = x_1 \begin{bmatrix} 1 & 0 \\ 0 & -1 \end{bmatrix} + x_2 \begin{bmatrix} 0 & 1 \\ 0 & 0 \end{bmatrix} + x_3 \begin{bmatrix} 0 & 0 \\ 1 & 0 \end{bmatrix}.$$

It follows that

$$\left\{ \begin{bmatrix} 1 & 0 \\ 0 & -1 \end{bmatrix}, \begin{bmatrix} 0 & 1 \\ 0 & 0 \end{bmatrix}, \begin{bmatrix} 0 & 0 \\ 1 & 0 \end{bmatrix} \right\}$$

spans W. It is easily verified that this set is linearly independent, and hence it is a basis for W.

33. T is linear but is not an isomorphism.

To prove that T is linear, let $f(x)$ and $g(x)$ be in $\mathcal{P}$. Then

$$T(f(x) + g(x)) = x(f(x) + g(x)) = xf(x) + xg(x) = T(f(x)) + T(g(x)),$$

and hence T preserves addition. Now let c be any scalar. Then

$$T((cf)(x)) = x((cf)(x)) = xcf(x) = c(xf(x)) = cT(f(x)),$$

and so T preserves scalar multiplication. Thus T is linear.

To show that T is not an isomorphism, we show that T is not onto. For any nonzero polynomial $f(x)$, $T(f(x)) = xf(x)$ has degree greater than zero and hence $T(f(x)) \neq 1$. Therefore the constant polynomial 1 is not in the range of T.

35. T is not linear. Let $f(x)$ be any nonzero polynomial. Then

$$T(2f(x)) = (2f(x))^2 = 4(f(x))^2 \neq 2(f(x))^2 = 2T(f(x)).$$

37. T is both linear and an isomorphism. Let f and g be in $\mathcal{F}(\mathcal{R})$. Then for any x in $\mathcal{R}$,

$$T(f+g)(x) = (f+g)(x+1) = f(x+1) + g(x+1) = T(f)(x) + T(g)(x),$$

and hence $T(f + g) = T(f) + T(g)$. Similarly, $T(cf) = cT(f)$ for any scalar c, and hence T is linear. To show that T is one-to-one, suppose that $T(f)$ is the zero function, that is, $T(f)(x) = f(x + 1) = 0$ for all x. Then for any x in $\mathcal{R}$, we have $f(x) = f((x - 1) + 1) = 0$, and hence f is the zero function. Thus T is one-to-one. To show T is onto, let f be any function in $\mathcal{F}(\mathcal{R})$ and g be the function in $\mathcal{F}(\mathcal{R})$ defined by $g(x) = f(x - 1)$ for all x in $\mathcal{R}$. Then for all x in $\mathcal{R}$, $T(g)(x) = g(x + 1) = f((x + 1) - 1) = f(x)$. Hence $T(g) = f$, and we conclude that T is onto.

41. Suppose that $\mathcal{B}$ is a basis for W_3. Since C_3 is not in W_3, the set $\mathcal{B} \cup \{C_3\}$ is linearly independent. Furthermore, this set spans V_3 as a consequence of the equation in Exercise 39(b). Hence it is a basis for V_3. Since this set contains 8 matrices, it follows that $\dim V_3 = 8$.

45. For each $i = 1, 2, \ldots, m$ and $j = 1, 2, \ldots, n$, let E_{ij} be the $m \times n$ matrix with a 1 as (i, j)-entry and zeros elsewhere. Since $\{E_{ij} : i = 1, 2, \ldots, m$ and $j = 1, 2, \ldots, n\}$ is a basis for $\mathcal{M}_{m \times n}$, the set $\{U(E_{ij}) : i = 1, 2, \ldots, m$ and $j = 1, 2, \ldots, n\}$, where U is the isomorphism defined in Example 12, is a basis for $\mathcal{L}(\mathcal{R}^n, \mathcal{R}^m)$.

49. (a) Since the zero polynomial can be written as $0x^4 + 0x^2 + 0$, it is in V. Let $p(x) = a_1 x^4 + b_1 x^2 + c_1$ and $q(x) = a_2 x^4 + b_2 x^2 + c_2$ be in V. Then $p(x) + q(x) = (a_1 + a_2)x^4 + (b_1 + b_2)x^2 + (c_1 + c_2)$, which is in V. Thus V is closed under addition. Similarly, V is closed under scalar multiplication, and hence V is a subspace of $\mathcal{P}_4$.

   (b) Let $T: V \to \mathcal{P}_2$ be defined by $T(ax^4 + bx^2 + c) = ax^2 + bx + c$. We show that T is linear. Let

   $$p(x) = a_1 x^4 + b_1 x^2 + c_1 \quad \text{and} \quad q(x) = a_2 x^4 + b_2 x^2 + c_2$$

   be in V. Then

   $$\begin{aligned} T(p(x) + q(x)) &= T((a_1 + a_2)x^4 + (b_1 + b_2)x^2 + (c_1 + c_2)) \\ &= (a_1 + a_2)x^2 + (b_1 + b_2)x + (c_1 + c_2) \\ &= (a_1 x^2 + b_1 x + c_1) + (a_2 x^2 + b_2 x + c_2) \\ &= T(p(x)) + T(q(x)). \end{aligned}$$

   Similarly, $T(cp(x)) = cT(p(x))$ for every $p(x)$ in V and every scalar c:

   $$\begin{aligned} T(cp(x)) &= T(c(a_1 x^4 + b_1 x^2 + c_1)) \\ &= T(ca_1 x^4 + cb_1 x^2 + cc_1) \\ &= ca_1 x^2 + cb_1 x + cc_1 \\ &= c(a_1 x^2 + b_1 x + c_1) \\ &= cT(p(x)). \end{aligned}$$

   Therefore T is linear.

To prove that T is one-to-one, suppose that $T(p(x)) = 0$, the zero polynomial. Since $a_1 x^2 + b_1 x + c_1 = 0$, it follows that $a_1 = b_1 = c_1 = 0$, and therefore $p(x) = a_1 x^4 + b_1 x^2 + c_1$ is the zero polynomial. Thus the null space of T is the zero subspace of V, and it follows that T is one-to-one.

53. Let $\mathcal{B}$ be a basis for V, and let $\Phi_{\mathcal{B}} \colon V \to \mathcal{R}^n$ be the isomorphism defined on page 411. Let $\mathcal{S}$ be a subset of V and $\mathcal{S}'$ denote the set of images of these vectors under $\Phi_{\mathcal{B}}$.

(a) Suppose that $\mathcal{S}$ spans V. Then $\mathcal{S}'$ spans $\mathcal{R}^n$, and hence contains at least n vectors. Therefore $\mathcal{S}$ contains at least n vectors.

(b) Suppose that $\mathcal{S}$ contains exactly n vectors. Then $\mathcal{S}'$ is a subset of $\mathcal{R}^n$ consisting of n linearly independent vectors, and hence is a basis for $\mathcal{R}^n$. Since $\Phi_{\mathcal{B}}^{-1}$ is an isomorphism (by Exercise 50) and $\mathcal{S}$ is the image of the vectors in $\mathcal{S}'$ under $\Phi_{\mathcal{B}}^{-1}$, it follows that $\mathcal{S}$ is a basis for V by Exercise 44.

57. We prove the following statement, which is equivalent to the given statement: If W is finite-dimensional, then V is finite-dimensional. Suppose that W is finite-dimensional, and let $\{\mathbf{v}_1, \mathbf{v}_2, \ldots, \mathbf{v}_n\}$ be a basis for W. Since W is isomorphic to V, there is an isomorphism $T \colon W \to V$. By Exercise 44, $\{T(\mathbf{v}_1), T(\mathbf{v}_2), \ldots, T(\mathbf{v}_n)\}$ is a basis for V, and hence V is finite-dimensional.

61. Suppose that $\dim V = n$ and $\dim W = m$. As an immediate consequence of Exercise 44, if V_1 and V_2 are isomorphic vector spaces and V_1 is finite-dimensional, then V_2 is finite-dimensional and $\dim V_1 = \dim V_2$. Thus it follows that $\dim \mathcal{L}(\mathcal{R}^n, \mathcal{R}^m) = \dim \mathcal{M}_{m \times n} = m \cdot n$ by Examples 11 and 12. So to establish the desired result, it suffices to show that $\mathcal{L}(\mathcal{R}^n, \mathcal{R}^m)$ is isomorphic to $\mathcal{L}(V, W)$.

Let $\mathcal{A}$ and $\mathcal{B}$ be bases for V and W, respectively. Let $\Phi_{\mathcal{A}}$ and $\Phi_{\mathcal{B}}$ be the isomorphisms from V to $\mathcal{R}^n$ and from W to $\mathcal{R}^m$ defined on page 411. Then $\Phi_{\mathcal{A}}^{-1}$ and $\Phi_{\mathcal{B}}^{-1}$ are isomorphisms by Exercise 50. For all T in $\mathcal{L}(\mathcal{R}^n, \mathcal{R}^m)$, define $\Phi \colon \mathcal{L}(\mathcal{R}^n, \mathcal{R}^m) \to \mathcal{L}(V, W)$ by

$$\Phi(T) = \Phi_{\mathcal{B}}^{-1} T \Phi_{\mathcal{A}}.$$

This definition makes sense because if T is a linear transformation from $\mathcal{R}^n$ to $\mathcal{R}^m$, then $\Phi(T) = \Phi_{\mathcal{B}}^{-1} T \Phi_{\mathcal{A}}$ is a linear transformation from V to W. We show that Φ is an isomorphism. First observe that Φ is linear. Let T_1 and T_2 be in $\mathcal{L}(\mathcal{R}^n, \mathcal{R}^m)$. Then for any $\mathbf{v}$ in V,

$$
\begin{aligned}
\Phi(T_1 + T_2)(\mathbf{v}) &= \Phi_{\mathcal{B}}^{-1}(T_1 + T_2)\Phi_{\mathcal{A}}(\mathbf{v}) \\
&= \Phi_{\mathcal{B}}^{-1}(T_1 \Phi_{\mathcal{A}}(\mathbf{v}) + T_2 \Phi_{\mathcal{A}}(\mathbf{v})) \\
&= \Phi_{\mathcal{B}}^{-1} T_1 \Phi_{\mathcal{A}}(\mathbf{v}) + \Phi_{\mathcal{B}}^{-1} T_2 \Phi_{\mathcal{A}}(\mathbf{v}) \\
&= \Phi(T_1)(\mathbf{v}) + \Phi(T_2)(\mathbf{v}) \\
&= (\Phi(T_1) + \Phi(T_2))(\mathbf{v}).
\end{aligned}
$$

Hence $\Phi(T_1 + T_2) = \Phi(T_1) + \Phi(T_2)$. Similarly, $\Phi(cT) = c\Phi(T)$ for any T in $\mathcal{L}(\mathcal{R}^n, \mathcal{R}^m)$ and any scalar c. Thus Φ is linear. To show that Φ is one-to-one and onto, we construct its inverse directly. Let $\Psi \colon \mathcal{L}(V, W) \to \mathcal{L}(\mathcal{R}^n, \mathcal{R}^m)$ be defined by

$$\Psi(T) = \Phi_{\mathcal{B}} T \Phi_{\mathcal{A}}^{-1}$$

for all T in $\mathcal{L}(V, W)$. Then for any T in $\mathcal{L}(\mathcal{R}^n, \mathcal{R}^m)$,

$$\Psi\Phi(T) = \Psi(\Phi_{\mathcal{B}}^{-1} T \Phi_{\mathcal{A}}) = \Phi_{\mathcal{B}}(\Phi_{\mathcal{B}}^{-1} T \Phi_{\mathcal{A}})\Phi_{\mathcal{A}}^{-1} = T.$$

Similarly, $\Phi\Psi(T) = T$ for all T in $\mathcal{L}(V, W)$. It follows that Φ is an isomorphism.

65. The set is linearly independent.

69. (rounded to 4 places after the decimal) $c_0 = 0.3486$, $c_1 = 0.8972$, $c_2 = -0.3667$, $c_3 = 0.1472$, $c_4 = -0.0264$

7.3 LINEAR TRANSFORMATIONS AND MATRIX REPRESENTATIONS

1. (a) False, let T be the $90°$-rotation operator on $\mathcal{R}^2$.
(b) False if D is the linear operator on the infinite-dimensional space C^∞ defined in Example 1.
(c) True
(d) True
(e) True
(f) True
(g) True

5. For any f and g in $\mathcal{F}(S)$,

$$T(f + g) = (f + g)(s) = f(s) + g(s) = T(f) + T(g).$$

For any scalar c and any f in $\mathcal{F}(S)$,

$$T(cf) = (cf)(s) = c(f(s)) = cT(f).$$

Therefore T is linear.

7. (a) For any A and B in $\mathcal{M}_{m \times n}$,

$$T(A + B) = (A + B)\mathbf{u} = A\mathbf{u} + B\mathbf{u} = T(A) + T(B).$$

Similarly, $T(cA) = cT(A)$ for any scalar c, and therefore T is linear.

(b) Let $\mathbf{v}$ be any vector in $\mathcal{R}^m$. Since $\mathbf{u} \neq \mathbf{0}$, there is an i such that $u_i \neq 0$. Let A be the $m \times n$ matrix whose ith column is $\frac{1}{u_i}\mathbf{v}$ and whose other columns are zero. Then

$$T(A) = A\mathbf{u}$$

$$= [\mathbf{0} \; \mathbf{0} \; \cdots \; \frac{1}{u_i}\mathbf{v} \; \cdots \; \mathbf{0}] \begin{bmatrix} u_1 \\ u_2 \\ \vdots \\ u_n \end{bmatrix}$$

$$= u_1 \mathbf{0} + u_2 \mathbf{0} + \cdots + u_i \frac{1}{u_i}\mathbf{v} + \cdots + u_n \mathbf{0}$$

$$= \mathbf{v}.$$

9. (a) $T(1) = (D^2 + D)(1) = 1'' + 1' = 0$ and

$$T(e^{-t}) = (D^2 + D)(e^{-t}) = (e^{-t})'' + (e^{-t})' = e^{-t} - e^{-t} = 0$$

(b) $T(e^{at}) = (D^2 + D)(e^{at}) = (e^{at})'' + (e^{at})' = a^2 e^{at} + a e^{at} = (a^2 + a)e^{at}$

13. Since

$$\begin{bmatrix} 1 & 2 \\ 3 & 4 \end{bmatrix} = 1 \begin{bmatrix} 1 & 0 \\ 0 & 0 \end{bmatrix} + 3 \begin{bmatrix} 0 & 0 \\ 1 & 0 \end{bmatrix} + 4 \begin{bmatrix} 0 & 0 \\ 0 & 1 \end{bmatrix} + 2 \begin{bmatrix} 0 & 1 \\ 0 & 0 \end{bmatrix},$$

it follows that

$$[A]_{\mathcal{B}} = \begin{bmatrix} 1 \\ 3 \\ 4 \\ 2 \end{bmatrix}.$$

17. Since

$$D(e^t) = 1e^t, \qquad D(e^{2t}) = 2e^{2t}, \qquad \text{and} \qquad D(e^{3t}) = 3e^{3t},$$

it follows that

$$[T]_{\mathcal{B}} = [D]_{\mathcal{B}} = \begin{bmatrix} 1 & 0 & 0 \\ 0 & 2 & 0 \\ 0 & 0 & 3 \end{bmatrix}.$$

21. (a) As in Example 10, let $\mathcal{B} = \{1, x, x^2\}$. Then

$$[D]_{\mathcal{B}} = \begin{bmatrix} 0 & 1 & 0 \\ 0 & 0 & 2 \\ 0 & 0 & 0 \end{bmatrix}.$$

So

$$[p'(x)]_{\mathcal{B}} = [D(p(x))]_{\mathcal{B}} = [D]_{\mathcal{B}}[p(x)]_{\mathcal{B}} = \begin{bmatrix} 0 & 1 & 0 \\ 0 & 0 & 2 \\ 0 & 0 & 0 \end{bmatrix} \begin{bmatrix} 6 \\ 0 \\ -4 \end{bmatrix} = \begin{bmatrix} 0 \\ -8 \\ 0 \end{bmatrix},$$

and hence $p'(x) = -8x$.

(b) Let $\mathcal{B}$ be as in (a). Then

$$[p'(x)]_{\mathcal{B}} = [D(p(x))]_{\mathcal{B}} = [D]_{\mathcal{B}}[p(x)]_{\mathcal{B}} = \begin{bmatrix} 0 & 1 & 0 \\ 0 & 0 & 2 \\ 0 & 0 & 0 \end{bmatrix} \begin{bmatrix} 2 \\ 3 \\ 5 \end{bmatrix} = \begin{bmatrix} 3 \\ 10 \\ 0 \end{bmatrix},$$

and hence $p'(x) = 3 + 10x$.

(c) Let $\mathcal{B} = \{1, x, x^2, x^3\}$. Then

$$[D]_{\mathcal{B}} = \begin{bmatrix} 0 & 1 & 0 & 0 \\ 0 & 0 & 2 & 0 \\ 0 & 0 & 0 & 3 \\ 0 & 0 & 0 & 0 \end{bmatrix}.$$

So

$$[p'(x)]_{\mathcal{B}} = [D(p(x))]_{\mathcal{B}} = [D]_{\mathcal{B}}[p(x)]_{\mathcal{B}} = \begin{bmatrix} 0 & 1 & 0 & 0 \\ 0 & 0 & 2 & 0 \\ 0 & 0 & 0 & 3 \\ 0 & 0 & 0 & 0 \end{bmatrix} \begin{bmatrix} 0 \\ 0 \\ 0 \\ 1 \end{bmatrix} = \begin{bmatrix} 0 \\ 0 \\ 3 \\ 0 \end{bmatrix},$$

and hence $p'(x) = 3x^2$.

25. Let $T = D$, the differential operator on $V = \text{Span}\{e^t, t^{2t}, e^{3t}\}$. By Exercise 17,

$$[T]_{\mathcal{B}} = [D]_{\mathcal{B}} = \begin{bmatrix} 1 & 0 & 0 \\ 0 & 2 & 0 \\ 0 & 0 & 3 \end{bmatrix},$$

and hence 1, 2, and 3 are the eigenvalues of D with corresponding bases $\{e^t\}$, $\{e^{2t}\}$, and $\{e^{3t}\}$.

29. (a) For any matrices A and C in $\mathcal{M}_{2\times 2}$,

$$T(A + C) = (\text{trace }(A + C))B = (\text{trace }A + \text{trace }C)B$$
$$= (\text{trace }A)B + (\text{trace }C)B = T(A) + T(C).$$

For any scalar c and any matrix A in $\mathcal{M}_{2\times 2}$,

$$T(cA) = (\text{trace }(cA))B = (c\,\text{trace }A)B = c((\text{trace }A)B) = cT(A).$$

Therefore T is linear.

(b) Let

$$\mathcal{B} = \left\{ \begin{bmatrix} 1 & 0 \\ 0 & 0 \end{bmatrix}, \begin{bmatrix} 0 & 1 \\ 0 & 0 \end{bmatrix}, \begin{bmatrix} 0 & 0 \\ 1 & 0 \end{bmatrix}, \begin{bmatrix} 0 & 0 \\ 0 & 1 \end{bmatrix} \right\}.$$

Then

$$T\left(\begin{bmatrix} 1 & 0 \\ 0 & 0 \end{bmatrix}\right) = \left(\text{trace}\begin{bmatrix} 1 & 0 \\ 0 & 0 \end{bmatrix}\right)\begin{bmatrix} 1 & 2 \\ 3 & 4 \end{bmatrix} = 1\begin{bmatrix} 1 & 2 \\ 3 & 4 \end{bmatrix}$$

$$= 1\begin{bmatrix} 1 & 0 \\ 0 & 0 \end{bmatrix} + 2\begin{bmatrix} 0 & 1 \\ 0 & 0 \end{bmatrix} + 3\begin{bmatrix} 0 & 0 \\ 1 & 0 \end{bmatrix} + 4\begin{bmatrix} 0 & 0 \\ 0 & 1 \end{bmatrix},$$

$$T\left(\begin{bmatrix} 0 & 1 \\ 0 & 0 \end{bmatrix}\right) = \left(\text{trace}\begin{bmatrix} 0 & 1 \\ 0 & 0 \end{bmatrix}\right)\begin{bmatrix} 1 & 2 \\ 3 & 4 \end{bmatrix} = 0\begin{bmatrix} 1 & 2 \\ 3 & 4 \end{bmatrix}$$

$$= \begin{bmatrix} 0 & 0 \\ 0 & 0 \end{bmatrix},$$

$$T\left(\begin{bmatrix} 0 & 0 \\ 1 & 0 \end{bmatrix}\right) = \left(\text{trace}\begin{bmatrix} 0 & 0 \\ 1 & 0 \end{bmatrix}\right)\begin{bmatrix} 1 & 2 \\ 3 & 4 \end{bmatrix} = 0\begin{bmatrix} 1 & 2 \\ 3 & 4 \end{bmatrix}$$

$$= \begin{bmatrix} 0 & 0 \\ 0 & 0 \end{bmatrix},$$

and

$$T\left(\begin{bmatrix} 0 & 0 \\ 0 & 1 \end{bmatrix}\right) = \left(\text{trace}\begin{bmatrix} 0 & 0 \\ 0 & 1 \end{bmatrix}\right)\begin{bmatrix} 1 & 2 \\ 3 & 4 \end{bmatrix} = 1\begin{bmatrix} 1 & 2 \\ 3 & 4 \end{bmatrix}$$

$$= 1\begin{bmatrix} 1 & 0 \\ 0 & 0 \end{bmatrix} + 2\begin{bmatrix} 0 & 1 \\ 0 & 0 \end{bmatrix} + 3\begin{bmatrix} 0 & 0 \\ 1 & 0 \end{bmatrix} + 4\begin{bmatrix} 0 & 0 \\ 0 & 1 \end{bmatrix}.$$

Hence

$$[T]_{\mathcal{B}} = \begin{bmatrix} 1 & 0 & 0 & 1 \\ 2 & 0 & 0 & 2 \\ 3 & 0 & 0 & 3 \\ 4 & 0 & 0 & 4 \end{bmatrix}.$$

(c) Suppose that A is a nonzero matrix with trace equal to zero. Then

$$T(A) = (\text{trace } A)B = 0B = O = 0A,$$

and hence A is an eigenvector of T with corresponding eigenvalue equal to 0.

(d) Suppose that A is an eigenvector of T with a corresponding nonzero eigenvalue λ. Then

$$\lambda A = T(A) = (\text{trace } A)B,$$

and hence $A = \left(\frac{\text{trace } A}{\lambda}\right) B$ because $\lambda \neq 0$.

33. Suppose that $c_1 e^{b_1 t} + c_2 e^{b_2 t} + \cdots + c_k e^{b_k t} = 0$ for all t. Apply T_{b_k} to both sides of this equation to obtain

$$c_1(b_1 - b_k)e^{b_1 t} + c_2(b_2 - b_k)e^{b_2 t} + \cdots + c_{k-1}(b_{k-1} - b_k)e^{b_{k-1} t} = 0$$

for all t. Suppose $k = 3$. Then as a consequence of Exercise 32, $c_1(b_1 - b_3) = c_2(b_2 - b_3) = 0$, and hence $c_1 = c_2 = 0$ because $b_1 - b_3 \neq 0$ and $b_2 - b_3 \neq 0$. Thus $c_3 e^{b_1 t} = 0$ for all t, and hence $c_3 = 0$. Thus Exercise 33 is established for $k = 3$. We can use this fact and apply the method above to establish Exercise 33 for $k = 4$. This approach can be continued to establish Exercise 33 for any k.

37. Suppose that $\mathcal{B} = \{\mathbf{v}_1, \mathbf{v}_2, \ldots, \mathbf{v}_n\}$. Then for $j = 1, 2, \ldots, n$, we have $[UT(\mathbf{v}_j)]_\mathcal{B} = [U]_\mathcal{B}[T(\mathbf{v}_j)]_\mathcal{B}$ by Theorem 7.6, and hence

$$[UT]_\mathcal{B} = [[UT(\mathbf{v}_1)]_\mathcal{B} \ [UT(\mathbf{v}_2)]_\mathcal{B} \ \cdots \ [UT(\mathbf{v}_n)]_\mathcal{B}]$$

$$= [[U]_\mathcal{B}[T(\mathbf{v}_1)]_\mathcal{B} \ [U]_\mathcal{B}[T(\mathbf{v}_2)]_\mathcal{B} \ \cdots \ [U]_\mathcal{B}[T(\mathbf{v}_n)]_\mathcal{B}]$$

$$= [U]_\mathcal{B}[[T(\mathbf{v}_1)]_\mathcal{B} \ [T(\mathbf{v}_2)]_\mathcal{B} \ \cdots \ [T(\mathbf{v}_n)]_\mathcal{B}]$$

$$= [U]_\mathcal{B}[T]_\mathcal{B}.$$

41. Let Y and Z denote the null space and the range of T, respectively, and suppose that $p = \dim Y$ and $q = \dim Z$. Let $\{\mathbf{y}_1, \mathbf{y}_2, \ldots, \mathbf{y}_p\}$ and $\{\mathbf{z}_1, \mathbf{z}_2, \ldots, \mathbf{z}_q\}$ be bases for Y and Z, respectively. For each i, choose a vector $\mathbf{v}_i$ in V such that $T(\mathbf{v}_i) = \mathbf{z}_i$. We prove that $\mathcal{B} = \{\mathbf{y}_1, \mathbf{y}_2, \ldots, \mathbf{y}_p, \mathbf{v}_1, \mathbf{v}_2, \ldots, \mathbf{v}_q\}$ is a basis for V. Notice that the vectors listed in $\mathcal{B}$ are distinct because $T(\mathbf{y}_i) = \mathbf{0}$ and $T(\mathbf{v}_j) = \mathbf{z}_j \neq \mathbf{0}$ for all i and j.

To prove that $\mathcal{B}$ is linearly independent, suppose that

$$a_1\mathbf{y}_1 + a_2\mathbf{y}_2 + \cdots + a_p\mathbf{y}_p + b_1\mathbf{v}_1 + b_2\mathbf{v}_2 + \cdots + b_q\mathbf{v}_q = \mathbf{0}.$$

Applying T to both sides of this equation yields

$$b_1\mathbf{z}_1 + b_2\mathbf{z}_2 + \cdots + b_q\mathbf{z}_q = \mathbf{0},$$

and hence $b_1 = b_2 = \cdots = b_q = 0$ because $\{\mathbf{z}_1, \mathbf{z}_2, \ldots, \mathbf{z}_q\}$ is linearly independent. Thus the original equation reduces to

$$a_1\mathbf{y}_1 + a_2\mathbf{y}_2 + \cdots + a_p\mathbf{y}_p = \mathbf{0},$$

and hence $a_1 = a_2 = \cdots = a_p = 0$ because $\{\mathbf{y}_1, \mathbf{y}_2, \ldots, \mathbf{y}_p\}$ is linearly independent. Since all of the coefficients in the original linear combination are zero, $\mathcal{B}$ is linearly independent.

To prove that $\mathcal{B}$ spans V, consider any vector $\mathbf{u}$ in V. Then $T(\mathbf{u})$ is in Z, and hence there are scalars $b_1, b_2, \ldots, b_q$ such that

$$
\begin{aligned}
T(\mathbf{u}) &= b_1 \mathbf{z}_1 + b_2 \mathbf{z}_2 + \cdots + b_q \mathbf{z}_q \\
&= b_1 T(\mathbf{v}_1) + b_2 T(\mathbf{v}_2) + \cdots + b_q T(\mathbf{v}_q) \\
&= T(b_1 \mathbf{v}_1 + b_2 \mathbf{v}_2 + \cdots + b_q \mathbf{v}_q).
\end{aligned}
$$

It follows that $\mathbf{u} - (b_1 \mathbf{v}_1 + b_2 \mathbf{v}_2 + \cdots + b_q \mathbf{v}_q)$ is in Y, and hence there are scalars $a_1, a_2, \ldots, a_p$ such that

$$
\mathbf{u} - (b_1 \mathbf{v}_1 + b_2 \mathbf{v}_2 + \cdots + b_q \mathbf{v}_q) = a_1 \mathbf{y}_1 + a_2 \mathbf{y}_2 + \cdots + a_p \mathbf{y}_p.
$$

So

$$
\mathbf{u} = a_1 \mathbf{y}_1 + a_2 \mathbf{y}_2 + \cdots + a_p \mathbf{y}_p + b_1 \mathbf{v}_1 + b_2 \mathbf{v}_2 + \cdots + b_q \mathbf{v}_q.
$$

Therefore $\mathcal{B}$ spans V, and hence is a basis for $\mathcal{B}$. Since $\mathcal{B}$ consists of $p + q$ vectors, it follows that

$$
\dim V = p + q = \dim Y + \dim Z.
$$

45. (rounded to 4 places after the decimal)

   (a) $-1.6533, 2.6277, 6.6533, 8.3723$

   (b) $\left\{ \begin{bmatrix} -0.1827 & -0.7905 \\ 0.5164 & 0.2740 \end{bmatrix}, \begin{bmatrix} 0.6799 & -0.4655 \\ -0.4655 & 0.3201 \end{bmatrix}, \begin{bmatrix} 0.4454 & 0.0772 \\ 0.5909 & -0.6681 \end{bmatrix}, \begin{bmatrix} 0.1730 & 0.3783 \\ 0.3783 & 0.8270 \end{bmatrix} \right\}$

   (c) $\begin{bmatrix} 0.2438a - 0.1736b + 0.0083c + 0.0496d & -0.2603a - 0.2893b + 0.3471c + 0.0826d \\ 0.0124a + 0.3471b - 0.0165c - 0.0992d & -0.1116a + 0.1240b - 0.1488c + 0.1074d \end{bmatrix}$

7.4   INNER PRODUCT SPACES

1. (a) False, it is a scalar.
   (b) True
   (c) False, it has scalar values.
   (d) False, any positive scalar multiple of an inner product is an inner product.
   (e) True
   (f) False, if the set contains the zero vector, it is linearly dependent.
   (g) True
   (h) True

(i) True

(j) True

(k) False, the indefinite integral of functions is not a scalar.

(l) True

3. We have

$$\langle f, g \rangle = \int_1^2 f(t)g(t)\, dt = \int_1^2 te^t\, dt$$

$$= (te^t - e^t)\big|_1^2 = (2e^2 - e^2) - (e - e) = e^2.$$

5. $\langle A, B \rangle = \text{trace } AB^T = \text{trace} \begin{bmatrix} 1 & -1 \\ 2 & 3 \end{bmatrix} \begin{bmatrix} 2 & 1 \\ 4 & 0 \end{bmatrix} = \text{trace} \begin{bmatrix} -2 & 1 \\ 16 & 2 \end{bmatrix} = 0.$

7. We have

$$\langle f(x), g(x) \rangle = \int_{-1}^1 (x^2 - 2)(3x + 5)\, dx = \int_{-1}^1 (3x^3 + 5x^2 - 6x - 10)\, dx$$

$$= \left(x^4 + \frac{5}{3}x^3 - 3x^2 - 10x \right)\Big|_{-1}^1 = \frac{10}{3} - 20 = -\frac{50}{3}.$$

9. Let $\mathbf{u}$, $\mathbf{v}$, and $\mathbf{w}$ be in $\mathcal{R}^n$. If $\mathbf{u} \neq \mathbf{0}$, then

$$\langle \mathbf{u}, \mathbf{u} \rangle = A\mathbf{u} \cdot \mathbf{u} = (A\mathbf{u})^T \mathbf{u} = \mathbf{u}^T A\mathbf{u} > 0$$

because A is positive definite, establishing axiom 1.

Notice that

$$\langle \mathbf{u}, \mathbf{v} \rangle = (A\mathbf{u}) \cdot \mathbf{v} = (A\mathbf{u})^T \mathbf{v} = \mathbf{u}^T A\mathbf{v} = \mathbf{u} \cdot (A\mathbf{v}) = (A\mathbf{v}) \cdot \mathbf{u} = \langle \mathbf{v}, \mathbf{u} \rangle,$$

establishing axiom 2.

Also

$$\langle \mathbf{u} + \mathbf{v}, \mathbf{w} \rangle = (A(\mathbf{u} + \mathbf{v})) \cdot \mathbf{w} = (A\mathbf{u} + A\mathbf{v}) \cdot \mathbf{w}$$

$$= (A\mathbf{u}) \cdot \mathbf{w} + (A\mathbf{v}) \cdot \mathbf{w} = \langle \mathbf{u}, \mathbf{w} \rangle + \langle \mathbf{u}, \mathbf{w} \rangle,$$

establishing axiom 3.

Finally, for any scalar a,

$$\langle a\mathbf{u}, \mathbf{v} \rangle = (A(a\mathbf{u})) \cdot \mathbf{v} = (a(A\mathbf{u})) \cdot \mathbf{v} = a((A\mathbf{u}) \cdot \mathbf{v}) = a\langle \mathbf{u}, \mathbf{v} \rangle,$$

establishing axiom 4.

11. Let f, g, and h be in $C([a, b])$.

Axiom 3 We have

$$\langle f + g, h \rangle = \int_a^b (f + g)(t)h(t)\, dt$$

$$= \int_a^b (f(t) + g(t))h(t)\, dt$$

$$= \int_a^b (f(t)h(t) + g(t)h(t))\, dt$$

$$= \int_a^b f(t)h(t)\, dt + \int_a^b g(t)h(t)\, dt$$

$$= \langle f, h \rangle + \langle g, h \rangle\,.$$

Axiom 4 For any scalar c, we have

$$\langle cf, g \rangle = \int_a^b (cf)(t)g(t)\, dt = \int_a^b (cf(t))g(t)\, dt = c\int_a^b f(t)g(t)\, dt = c\,\langle f, g \rangle\,.$$

13. Yes. We verify the axioms of an inner product.

Axiom 1 Let $\mathbf{u}$ be a nonzero vector in V. Then $\langle \mathbf{u}, \mathbf{u} \rangle = 2(\mathbf{u} \cdot \mathbf{u}) \neq 0$ because $\mathbf{u} \cdot \mathbf{u} \neq 0$.

Axiom 2 Let $\mathbf{u}$ and $\mathbf{v}$ be in V. Then

$$\langle \mathbf{u}, \mathbf{v} \rangle = 2(\mathbf{u} \cdot \mathbf{v}) = 2(\mathbf{v} \cdot \mathbf{u}) = \langle \mathbf{v}, \mathbf{u} \rangle\,.$$

Axiom 3 Let $\mathbf{u}$, $\mathbf{v}$, and $\mathbf{w}$ be in V. Then

$$\langle \mathbf{u} + \mathbf{v}, \mathbf{w} \rangle = 2((\mathbf{u} + \mathbf{v}) \cdot \mathbf{w}) = 2(\mathbf{u} \cdot \mathbf{w} + \mathbf{v} \cdot \mathbf{w})$$
$$= 2(\mathbf{u} \cdot \mathbf{w}) + 2(\mathbf{v} \cdot \mathbf{w}) = \langle \mathbf{u}, \mathbf{w} \rangle + \langle \mathbf{v}, \mathbf{w} \rangle\,.$$

Axiom 4 Let $\mathbf{u}$ and $\mathbf{v}$ be in V, and let a be a scalar. Then

$$\langle a\mathbf{u}, \mathbf{v} \rangle = 2(a\mathbf{u} \cdot \mathbf{v}) = 2a(\mathbf{u} \cdot \mathbf{v}) = a(2(\mathbf{u} \cdot \mathbf{v})) = a\,\langle \mathbf{u}, \mathbf{v} \rangle\,.$$

15. No. We show that axiom 1 is not satisfied. Let $f\colon [0, 2] \to \mathcal{R}$ be defined by

$$f(t) = \begin{cases} 0 & \text{if } 0 \leq t \leq 1 \\ t - 1 & \text{if } 1 < t \leq 2. \end{cases}$$

Since f is continuous, it is in V. Furthermore, f is not the zero function. However

$$\langle f, f \rangle = \int_0^1 [f(t)]^2\, dt = \int_0^1 0\, dt = 0.$$

17. Yes. We verify the axioms of an inner product.

Axiom 1   Let $\mathbf{u}$ be a nonzero vector in V. Then

$$\langle \mathbf{u}, \mathbf{u} \rangle = \langle \mathbf{u}, \mathbf{u} \rangle_1 + \langle \mathbf{u}, \mathbf{u} \rangle_2 > 0$$

since $\langle \mathbf{u}, \mathbf{u} \rangle_1 > 0$ and $\langle \mathbf{u}, \mathbf{u} \rangle_2 > 0$.

Axiom 2   Let $\mathbf{u}$ and $\mathbf{v}$ be in V. Then

$$\langle \mathbf{u}, \mathbf{v} \rangle = \langle \mathbf{u}, \mathbf{v} \rangle_1 + \langle \mathbf{u}, \mathbf{v} \rangle_2 = \langle \mathbf{v}, \mathbf{u} \rangle_1 + \langle \mathbf{v}, \mathbf{u} \rangle_2 = \langle \mathbf{v}, \mathbf{u} \rangle .$$

Axiom 3   Let $\mathbf{u}$, $\mathbf{v}$, and $\mathbf{w}$ be in V. Then

$$\begin{aligned}
\langle \mathbf{u} + \mathbf{v}, \mathbf{w} \rangle &= \langle \mathbf{u} + \mathbf{v}, \mathbf{w} \rangle_1 + \langle \mathbf{u} + \mathbf{v}, \mathbf{w} \rangle_2 \\
&= \langle \mathbf{u}, \mathbf{w} \rangle_1 + \langle \mathbf{v}, \mathbf{w} \rangle_1 + \langle \mathbf{u}, \mathbf{w} \rangle_2 + \langle \mathbf{v}, \mathbf{w} \rangle_2 \\
&= \langle \mathbf{u}, \mathbf{w} \rangle_1 + \langle \mathbf{u}, \mathbf{w} \rangle_2 + \langle \mathbf{v}, \mathbf{w} \rangle_1 + \langle \mathbf{v}, \mathbf{w} \rangle_2 \\
&= \langle \mathbf{u}, \mathbf{w} \rangle + \langle \mathbf{v}, \mathbf{w} \rangle .
\end{aligned}$$

Axiom 4   Let $\mathbf{u}$ and $\mathbf{v}$ be in V, and let a be a scalar. Then

$$\begin{aligned}
\langle a\mathbf{u}, \mathbf{v} \rangle &= \langle a\mathbf{u}, \mathbf{v} \rangle_1 + \langle a\mathbf{u}, \mathbf{v} \rangle_2 \\
&= a \langle \mathbf{u}, \mathbf{v} \rangle_1 + a \langle \mathbf{u}, \mathbf{v} \rangle_2 \\
&= a(\langle \mathbf{u}, \mathbf{v} \rangle_1 + \langle \mathbf{u}, \mathbf{v} \rangle_2) \\
&= a \langle \mathbf{u}, \mathbf{v} \rangle .
\end{aligned}$$

21. Let $\mathbf{u}_1 = 1$, $\mathbf{u}_2 = e^t$, and $\mathbf{u}_3 = e^{-t}$. We apply the Gram-Schmidt process to $\{\mathbf{u}_1, \mathbf{u}_2, \mathbf{u}_3\}$ to obtain an orthogonal basis $\{\mathbf{v}_1, \mathbf{v}_2, \mathbf{v}_3\}$. Let $\mathbf{v}_1 = \mathbf{u}_1 = 1$,

$$\begin{aligned}
\mathbf{v}_2 &= \mathbf{u}_2 - \frac{\langle \mathbf{u}_2, \mathbf{v}_1 \rangle}{\|\mathbf{v}_1\|^2} \mathbf{v}_1 \\[2mm]
&= e^t - \frac{\int_0^1 e^t 1\, dt}{\int_0^1 1^2\, dt} 1 \\[2mm]
&= e^t - \frac{e-1}{1} 1 = e^t - e + 1,
\end{aligned}$$

and

$$\begin{aligned}
\mathbf{v}_3 &= \mathbf{u}_3 - \frac{\langle \mathbf{u}_3, \mathbf{v}_1 \rangle}{\|\mathbf{v}_1\|^2} \mathbf{v}_1 - \frac{\langle \mathbf{u}_3, \mathbf{v}_2 \rangle}{\|\mathbf{v}_2\|^2} \mathbf{v}_2 \\[2mm]
&= e^{-t} - \frac{\int_0^1 e^{-t} 1\, dt}{\int_0^1 1^2\, dt} 1 - \frac{\int_0^1 e^{-t}(e^t - e + 1)\, dt}{\int_0^1 (e^t - e + 1)^2\, dt}(e^t - e + 1) \\[2mm]
&= e^{-t} - \frac{e-1}{e} - \frac{2(e^2 - 3e + 1)}{e(e-1)(e-3)}(e^t - e + 1)
\end{aligned}$$

$$= e^{-t} + \frac{e^2 - 2e - 1}{e(e-3)} - \frac{2(e^2 - 3e + 1)}{e(e-1)(e-3)} e^t.$$

Thus

$$\{v_1, v_2, v_3\} = \left\{ 1, e^t - e + 1, e^{-t} + \frac{e^2 - 2e - 1}{e(e-3)} - \frac{2(e^2 - 3e + 1)}{e(e-1)(e-3)} e^t \right\}.$$

25. We have $\langle \mathbf{u}, \mathbf{0} \rangle = \langle \mathbf{0}, \mathbf{u} \rangle = \langle 0\mathbf{0}, \mathbf{u} \rangle = 0 \langle \mathbf{0}, \mathbf{u} \rangle = 0.$

29. Suppose that $\langle \mathbf{u}, \mathbf{w} \rangle = 0$ for all $\mathbf{u}$ in V. Since $\mathbf{w}$ is in V, we have $\langle \mathbf{w}, \mathbf{w} \rangle = 0$, and hence $\mathbf{w} = \mathbf{0}$ by axiom 1.

33. Observe that

$$AB^T = \begin{bmatrix} a_{11} & a_{12} \\ a_{21} & a_{22} \end{bmatrix} \begin{bmatrix} b_{11} & b_{21} \\ b_{12} & b_{22} \end{bmatrix} = \begin{bmatrix} a_{11}b_{11} + a_{12}b_{12} & a_{11}b_{21} + a_{12}b_{22} \\ a_{21}b_{11} + a_{22}b_{12} & a_{21}b_{21} + a_{22}b_{22} \end{bmatrix},$$

and hence $\langle A, B \rangle = \text{trace}\,(AB^T) = a_{11}b_{11} + a_{12}b_{12} + a_{21}b_{21} + a_{22}b_{22}.$

37. If $\mathbf{u}$ or $\mathbf{v}$ is the zero vector, then both sides of the equality have the value zero. So suppose that $\mathbf{u} \neq \mathbf{0}$ and $\mathbf{v} \neq \mathbf{0}$. Then there exists a scalar c such that $\mathbf{v} = c\mathbf{u}$. Hence $\langle \mathbf{u}, \mathbf{v} \rangle^2 = \langle \mathbf{u}, c\mathbf{u} \rangle^2 = c^2 \langle \mathbf{u}, \mathbf{u} \rangle^2$, and

$$\langle \mathbf{u}, \mathbf{u} \rangle \langle \mathbf{v}, \mathbf{v} \rangle = \langle \mathbf{u}, \mathbf{u} \rangle \langle c\mathbf{u}, c\mathbf{u} \rangle = \langle \mathbf{u}, \mathbf{u} \rangle c^2 \langle \mathbf{u}, \mathbf{u} \rangle = c^2 \langle \mathbf{u}, \mathbf{u} \rangle^2.$$

Therefore $\langle \mathbf{u}, \mathbf{v} \rangle^2 = \langle \mathbf{u}, \mathbf{u} \rangle \langle \mathbf{v}, \mathbf{v} \rangle.$

41. (a) By the argument on page 90, $B^T B$ and $B B^T$ are symmetric. Suppose that B is invertible. Let $\mathbf{v}$ be any nonzero vector in $\mathcal{R}^n$. Then $B\mathbf{v} \neq \mathbf{0}$ by Theorem 2.6 on page 126. Thus

$$\mathbf{v}^T B^T B \mathbf{v} = (B\mathbf{v})^T (B\mathbf{v}) = (B\mathbf{v}) \cdot (B\mathbf{v}) > 0,$$

and hence $B^T B$ is positive definite.

(b) Let $\mathcal{B}$ be a basis for $\mathcal{R}^n$ that is orthonormal with respect to the given inner product, let B be the $n \times n$ matrix whose columns are the vectors in $\mathcal{B}$, and let $A = (B^{-1})^T B^{-1}$. (Although $\mathcal{B}$ is orthonormal with respect to the given inner product, it need not be orthonormal with respect to the usual dot product on $\mathcal{R}^n$.) Then A is positive definite by (a). Furthermore, by Theorem 4.9, $[\mathbf{u}]_{\mathcal{B}} = B^{-1}\mathbf{u}$ for any vector $\mathbf{u}$ in $\mathcal{R}^n$. So, for any vectors $\mathbf{u}$ and $\mathbf{v}$ in $\mathcal{R}^n$, we may apply Exercise 31 to obtain

$$\langle \mathbf{u}, \mathbf{v} \rangle = [\mathbf{u}]_{\mathcal{B}} \cdot [\mathbf{v}]_{\mathcal{B}} = (B^{-1}\mathbf{u}) \cdot (B^{-1}\mathbf{v}) = (B^{-1}\mathbf{u})^T (B^{-1}\mathbf{v})$$
$$= \mathbf{u}^T (B^{-1})^T (B^{-1}\mathbf{v}) = \mathbf{u}^T A\mathbf{v} = (A\mathbf{u})^T \mathbf{v} = (A\mathbf{u}) \cdot \mathbf{v}.$$

45. Let $\mathbf{u}$ and $\mathbf{v}$ be in W. Then by Exercise 44,

$$\mathbf{u} + \mathbf{v} = \langle \mathbf{u} + \mathbf{v}, \mathbf{w}_1 \rangle \mathbf{w}_1 + \langle \mathbf{u} + \mathbf{v}, \mathbf{w}_2 \rangle \mathbf{w}_2 + \cdots + \langle \mathbf{u} + \mathbf{v}, \mathbf{w}_n \rangle \mathbf{w}_n$$
$$= (\langle \mathbf{u}, \mathbf{w}_1 \rangle + \langle \mathbf{v}, \mathbf{w}_1 \rangle)\mathbf{w}_1 + \cdots + (\langle \mathbf{u}, \mathbf{w}_n \rangle + \langle \mathbf{v}, \mathbf{w}_n \rangle)\mathbf{w}_n.$$

CHAPTER 7 REVIEW

1. (a) False, for example, C^∞ is not a subset of $\mathcal{R}^n$ for any n.
   (b) True
   (c) False, the dimension is mn.
   (d) False, it is an $mn \times mn$ matrix.
   (e) True
   (f) False, for example, let $\mathbf{u}$ and $\mathbf{w}$ be any vectors in an inner product space that are not orthogonal, and let $\mathbf{v} = \mathbf{0}$.
   (g) True

3. No. First observe that $a \oplus 0 = a + 0 + a \cdot 0 = a$ for all a in V, and hence 0 is the (necessarily unique) additive identity for V. However, for any v in V,

$$(-1) \oplus v = (-1) + v + (-1)v = -1 \neq 0,$$

and hence -1 has no additive inverse, so that axiom 4 fails.

5. Yes. We verify a few of the axioms of a vector space. The others are left to the reader.
   Axiom 2 Let f, g, and h be in V. Then for any x in $\mathcal{R}$,

$$((f \oplus g) \oplus h)(x) = (f \oplus g)(x)h(x) = (f(x)g(x))h(x) = f(x)(g(x)h(x))$$
$$= f(x)(g \oplus h)(x) = (f \oplus (g \oplus h))(x),$$

and therefore $(f \oplus g) \oplus h = f \oplus (g \oplus h)$.
   Axiom 7 Let f and g be in V, and let a be a scalar. Then for any x in $\mathcal{R}$,

$$(a \odot (f \oplus g))(x) = ((f \oplus g)(x))^a = (f(x)g(x))^a = f(x)^a g(x)^a$$
$$= ((a \odot f)(x))((a \odot g)(x)) = ((a \odot f) \oplus (a \odot g))(x),$$

and hence $a \odot (f \oplus g) = (a \odot f) \oplus (a \odot g)$.

7. No. V is not closed under addition. For example, although $x - x^2$ and $x + x^2$ are in V, their sum $(x - x^2) + (x + x^2) = 2x$ is not in V.

9. No. Since $\lambda \neq 0$, it follows that λ is not an eigenvalue of O, and hence O is not in W. Therefore W is not a subspace of V.

11. No. Consider the matrix equation

$$x_1 \begin{bmatrix} 1 & 2 \\ 1 & -1 \end{bmatrix} + x_2 \begin{bmatrix} 0 & 1 \\ 2 & 0 \end{bmatrix} + x_3 \begin{bmatrix} -1 & 3 \\ 1 & 1 \end{bmatrix} = \begin{bmatrix} 2 & 8 \\ 1 & -5 \end{bmatrix}.$$

Comparing the $(1, 1)$-entries of both sides of this equation, we obtain $x_1 - x_3 = 2$. Comparing the $(2, 2)$-entries of both sides of this equation, we obtain $-x_1 + x_3 = 5$, which is equivalent to $x_1 - x_3 = -5$. Since the system consisting of these two equations is inconsistent, the matrix equation has no solution, and hence the given matrix is not a linear combination of the matrices in the given set.

13. Yes. Consider the matrix equation

$$x_1 \begin{bmatrix} 1 & 2 \\ 1 & -1 \end{bmatrix} + x_2 \begin{bmatrix} 0 & 1 \\ 2 & 0 \end{bmatrix} + x_3 \begin{bmatrix} -1 & 3 \\ 1 & 1 \end{bmatrix} = \begin{bmatrix} 4 & 1 \\ -2 & -4 \end{bmatrix}.$$

Comparing the corresponding entries on both sides of this equation, we obtain the system

$$\begin{aligned}
x_1 \quad\quad - \ x_3 &= \ \ 4 \\
2x_1 + \ x_2 + 3x_3 &= \ \ 1 \\
x_1 + 2x_2 + \ x_3 &= -2 \\
-x_1 \quad\quad + \ x_3 &= -4
\end{aligned}$$

whose coefficient matrix has the reduced row echelon form

$$\begin{bmatrix} 1 & 0 & 0 & 3 \\ 0 & 1 & 0 & -2 \\ 0 & 0 & 1 & -1 \\ 0 & 0 & 0 & 0 \end{bmatrix}.$$

Therefore the system has the solution $x_1 = 3$, $x_2 = -2$, and $x_3 = -1$. These are the coefficients of the linear combination to produce the given matrix.

17. A polynomial $f(x) = a + bx + cx^2 + dx^3$ is in W if and only if

$$f(0) + f'(0) + f''(0) = a + b + 2c = 0, \quad\quad \text{or} \quad\quad a = -b - 2c.$$

So $f(x)$ is in W if and only if

$$\begin{aligned}
f(x) &= (-b - 2c) + bx + cx^2 + dx^3 \\
&= b(-1 + x) + c(-2 + x^2) + dx^3.
\end{aligned}$$

It follows that W is the span of $\mathcal{B} = \{-1 + x, -2 + x^2, x^3\}$.

Next, we show that $\mathcal{B}$ is linearly independent. Suppose that

$$a(-1 + x) + b(-2 + x^2) + cx^3 = \mathbf{0},$$

where $\mathbf{0}$ is the zero polynomial. This equation can be rewritten

$$(-a - 2b) + ax + bx^2 + cx^3 = \mathbf{0}$$

from which it follows that

$$\begin{aligned}
-a - 2b \quad\quad &= 0 \\
a \quad\quad\quad &= 0 \\
b \quad &= 0 \\
c &= 0.
\end{aligned}$$

Therefore $a = b = c = 0$, and hence $\mathcal{B}$ is linearly independent. It follows that $\mathcal{B}$ is a basis for W, and therefore $\dim W = 3$.

19. T is not linear. For example, let $A = I_2$. Then $T(2A) = \text{trace}\,(4I_2) = 8$, but $2T(A) = 2\,\text{trace}\,(I_2) = 4$.

21. T is both linear and an isomorphism. Let $f(x)$ and $g(x)$ be in $\mathcal{P}_2$. Then

$$T(f(x) + g(x)) = \begin{bmatrix} (f+g)(0) \\ (f+g)'(0) \\ \int_0^1 (f+g)(t)\,dt \end{bmatrix}$$

$$= \begin{bmatrix} f(0) + g(0) \\ f'(0) + g'(0) \\ \int_0^1 f(t)\,dt + \int_0^1 g(t)\,dt \end{bmatrix}$$

$$= \begin{bmatrix} f(0) \\ f'(0) \\ \int_0^1 f(t)\,dt \end{bmatrix} + \begin{bmatrix} g(0) \\ g'(0) \\ \int_0^1 g(t)\,dt \end{bmatrix}$$

$$= T(f(x)) + T(g(x)).$$

Thus T preserves addition. Furthermore, for any scalar c,

$$T(cf(x)) = \begin{bmatrix} (cf)(0) \\ (cf)'(0) \\ \int_0^1 cf(t)\,dt \end{bmatrix} = \begin{bmatrix} cf(0) \\ cf'(0) \\ c\int_0^1 f(t)\,dt \end{bmatrix} = c \begin{bmatrix} f(0) \\ f'(0) \\ \int_0^1 f(t)\,dt \end{bmatrix} = cT(f(x)),$$

and hence T preserves scalar multiplication. Therefore T is linear.

To show that T is an isomorphism, it suffices to show that T is one-to-one because the domain and the codomain of T are finite-dimensional vector spaces with the same dimension. Suppose $f(x) = a + bx + cx^2$ is a polynomial in $\mathcal{P}_2$ such that $T(f(x)) = \mathbf{0}$, the zero polynomial. Comparing components in this vector equation, we have

$$f(0) = 0, \qquad f'(0) = 0, \qquad \text{and} \qquad \int_0^1 f(t)dt = 0.$$

Since $f(0) = a + b0 + c0^2 = a$, we have $a = 0$. Similarly, we obtain that $b = 0$ from the second equation, and hence $c = 0$ follows from the third equation. Therefore f is the zero polynomial, and so the null space of T is the zero subspace. We conclude that T is one-to-one, and hence T is an isomorphism.

25. We have

$$T\left(\begin{bmatrix} 1 & 0 \\ 0 & 0 \end{bmatrix}\right) = 2\begin{bmatrix} 1 & 0 \\ 0 & 0 \end{bmatrix} + \begin{bmatrix} 1 & 0 \\ 0 & 0 \end{bmatrix}^T = 3\begin{bmatrix} 1 & 0 \\ 0 & 0 \end{bmatrix},$$

$$T\left(\begin{bmatrix} 0 & 1 \\ 0 & 0 \end{bmatrix}\right) = 2\begin{bmatrix} 0 & 1 \\ 0 & 0 \end{bmatrix} + \begin{bmatrix} 0 & 1 \\ 0 & 0 \end{bmatrix}^T = 2\begin{bmatrix} 0 & 1 \\ 0 & 0 \end{bmatrix} + \begin{bmatrix} 0 & 0 \\ 1 & 0 \end{bmatrix},$$

$$T\left(\begin{bmatrix} 0 & 0 \\ 1 & 0 \end{bmatrix}\right) = 2\begin{bmatrix} 0 & 0 \\ 1 & 0 \end{bmatrix} + \begin{bmatrix} 0 & 0 \\ 1 & 0 \end{bmatrix}^T = 2\begin{bmatrix} 0 & 0 \\ 1 & 0 \end{bmatrix} + \begin{bmatrix} 0 & 1 \\ 0 & 0 \end{bmatrix},$$

and

$$T\left(\begin{bmatrix} 0 & 0 \\ 0 & 1 \end{bmatrix}\right) = 2\begin{bmatrix} 0 & 0 \\ 0 & 1 \end{bmatrix} + \begin{bmatrix} 0 & 0 \\ 0 & 1 \end{bmatrix}^T = 3\begin{bmatrix} 0 & 0 \\ 0 & 1 \end{bmatrix}.$$

Therefore

$$[T]_B = \begin{bmatrix} 3 & 0 & 0 & 0 \\ 0 & 2 & 1 & 0 \\ 0 & 1 & 2 & 0 \\ 0 & 0 & 0 & 3 \end{bmatrix}.$$

29. Using the matrix computed in Exercise 25, we have

$$[T^{-1}]_B = \begin{bmatrix} 3 & 0 & 0 & 0 \\ 0 & 2 & 1 & 0 \\ 0 & 1 & 2 & 0 \\ 0 & 0 & 0 & 3 \end{bmatrix}^{-1} = \frac{1}{3}\begin{bmatrix} 1 & 0 & 0 & 0 \\ 0 & 2 & -1 & 0 \\ 0 & -1 & 2 & 0 \\ 0 & 0 & 0 & 1 \end{bmatrix}.$$

Hence for any matrix $\begin{bmatrix} a & b \\ c & d \end{bmatrix}$ in $\mathcal{M}_{2\times 2}$,

$$\left[T^{-1}\left(\begin{bmatrix} a & b \\ c & d \end{bmatrix}\right)\right]_B = [T^{-1}]_B\left[\begin{bmatrix} a & b \\ c & d \end{bmatrix}\right]_B$$

$$= \frac{1}{3}\begin{bmatrix} 1 & 0 & 0 & 0 \\ 0 & 2 & -1 & 0 \\ 0 & -1 & 2 & 0 \\ 0 & 0 & 0 & 1 \end{bmatrix}\begin{bmatrix} a \\ b \\ c \\ d \end{bmatrix}.$$

$$= \frac{1}{3}\begin{bmatrix} a \\ 2b - c \\ -b + 2c \\ d \end{bmatrix},$$

and hence

$$T^{-1}\left(\begin{bmatrix} a & b \\ c & d \end{bmatrix}\right) = \frac{1}{3}\begin{bmatrix} a & 2b - c \\ -b + 2c & d \end{bmatrix}.$$

33. Let $A = [T]_\mathcal{B}$. By Exercise 25,

$$A = \begin{bmatrix} 3 & 0 & 0 & 0 \\ 0 & 2 & 1 & 0 \\ 0 & 1 & 2 & 0 \\ 0 & 0 & 0 & 3 \end{bmatrix}.$$

We first find the eigenvalues of A. The characteristic polynomial of A is

$$\det(A - tI_4) = \det \begin{bmatrix} 3-t & 0 & 0 & 0 \\ 0 & 2-t & 1 & 0 \\ 0 & 1 & 2-t & 0 \\ 0 & 0 & 0 & 3-t \end{bmatrix} = (t-3)^3(t-1),$$

and hence the eigenvalues of A are 3 and 1.

Next, we find a basis for the eigenspace of A corresponding to the eigenvalue 3. This eigenspace is the solution space of the system of equations

$$(A - 3I_4)\mathbf{x} = \begin{bmatrix} 0 & 0 & 0 & 0 \\ 0 & -1 & 1 & 0 \\ 0 & 1 & -1 & 0 \\ 0 & 0 & 0 & 0 \end{bmatrix} \begin{bmatrix} x_1 \\ x_2 \\ x_3 \\ x_4 \end{bmatrix} = \begin{bmatrix} 0 \\ 0 \\ 0 \\ 0 \end{bmatrix}.$$

Hence $x_2 = x_3$, and the parametric representation of the general solution is

$$\begin{bmatrix} x_1 \\ x_2 \\ x_3 \\ x_4 \end{bmatrix} = x_1 \begin{bmatrix} 1 \\ 0 \\ 0 \\ 0 \end{bmatrix} + x_2 \begin{bmatrix} 0 \\ 1 \\ 1 \\ 0 \end{bmatrix} + x_4 \begin{bmatrix} 0 \\ 0 \\ 0 \\ 1 \end{bmatrix}.$$

Thus

$$\left\{ \begin{bmatrix} 1 \\ 0 \\ 0 \\ 0 \end{bmatrix}, \begin{bmatrix} 0 \\ 1 \\ 1 \\ 0 \end{bmatrix}, \begin{bmatrix} 0 \\ 0 \\ 0 \\ 1 \end{bmatrix} \right\}$$

is a basis for the eigenspace of A corresponding to the eigenvalue 3. Let A_1, A_2, and A_3 be the matrices whose coordinate vectors relative to $\mathcal{B}$ are the vectors in the basis above. Then

$$\{A_1, A_2, A_3\} = \left\{ \begin{bmatrix} 1 & 0 \\ 0 & 0 \end{bmatrix}, \begin{bmatrix} 0 & 1 \\ 1 & 0 \end{bmatrix}, \begin{bmatrix} 0 & 0 \\ 0 & 1 \end{bmatrix} \right\}$$

is a basis for the eigenspace of T corresponding to the eigenvalue 3.

Finally, we find a basis for the eigenspace of A corresponding to the eigenvalue 1. This eigenspace is the solution space of the system of equations

$$(A - I_4)\mathbf{x} = \begin{bmatrix} 2 & 0 & 0 & 0 \\ 0 & 1 & 1 & 0 \\ 0 & 1 & 1 & 0 \\ 0 & 0 & 0 & 2 \end{bmatrix} \begin{bmatrix} x_1 \\ x_2 \\ x_3 \\ x_4 \end{bmatrix} = \begin{bmatrix} 0 \\ 0 \\ 0 \\ 0 \end{bmatrix}.$$

Thus $x_1 = 0$, $x_3 = -x_2$, and $x_4 = 0$. So the parametric representation of the general solution is

$$\begin{bmatrix} x_1 \\ x_2 \\ x_3 \\ x_4 \end{bmatrix} = x_2 \begin{bmatrix} 0 \\ 1 \\ -1 \\ 0 \end{bmatrix}.$$

Let A_4 be the matrix such that

$$[A_4]_\mathcal{B} = \begin{bmatrix} 0 \\ 1 \\ -1 \\ 0 \end{bmatrix}.$$

Then as in the case for eigenvalue 3, we have that

$$\{A_4\} = \left\{ \begin{bmatrix} 0 & 1 \\ -1 & 0 \end{bmatrix} \right\}$$

is a basis for the eigenspace of T corresponding to the eigenvalue 1.

37. For any matrix $\begin{bmatrix} a & b \\ c & d \end{bmatrix}$ in $\mathcal{M}_{2\times 2}$,

$$\operatorname{trace}\left(\begin{bmatrix} 0 & 1 \\ 1 & 0 \end{bmatrix} \begin{bmatrix} a & b \\ c & d \end{bmatrix} \right) = \operatorname{trace} \begin{bmatrix} c & d \\ a & b \end{bmatrix} = c + b,$$

and hence the matrix is in W if and only if $c = -b$. Thus the matrix is in W if and only if it is of the form

$$\begin{bmatrix} a & b \\ -b & d \end{bmatrix} = a \begin{bmatrix} 1 & 0 \\ 0 & 0 \end{bmatrix} + b \begin{bmatrix} 0 & 1 \\ -1 & 0 \end{bmatrix} + d \begin{bmatrix} 0 & 0 \\ 0 & 1 \end{bmatrix}.$$

Observe that the vectors in this linear combination form an orthogonal set. We divide each such vector by its length to obtain the orthonormal basis

$$\{M_1, M_2, M_3\} = \left\{ \begin{bmatrix} 1 & 0 \\ 0 & 0 \end{bmatrix}, \frac{1}{\sqrt{2}} \begin{bmatrix} 0 & 1 \\ -1 & 0 \end{bmatrix}, \begin{bmatrix} 0 & 0 \\ 0 & 1 \end{bmatrix} \right\}$$

for W. Thus the orthogonal projection B of $A = \begin{bmatrix} 2 & 5 \\ 9 & -3 \end{bmatrix}$ on W is

$$B = \langle M_1, A \rangle\, M_1 + \langle M_2, A \rangle\, M_2 + \langle M_3, A \rangle\, M_3$$

$$= \left\langle \begin{bmatrix} 1 & 0 \\ 0 & 0 \end{bmatrix}, \begin{bmatrix} 2 & 5 \\ 9 & -3 \end{bmatrix} \right\rangle \begin{bmatrix} 1 & 0 \\ 0 & 0 \end{bmatrix} + \left\langle \frac{1}{\sqrt{2}} \begin{bmatrix} 0 & 1 \\ -1 & 0 \end{bmatrix}, \begin{bmatrix} 2 & 5 \\ 9 & -3 \end{bmatrix} \right\rangle \frac{1}{\sqrt{2}} \begin{bmatrix} 0 & 1 \\ -1 & 0 \end{bmatrix}$$

$$+ \left\langle \begin{bmatrix} 0 & 0 \\ 0 & 1 \end{bmatrix}, \begin{bmatrix} 2 & 5 \\ 9 & -3 \end{bmatrix} \right\rangle \begin{bmatrix} 0 & 0 \\ 0 & 1 \end{bmatrix}$$

$$= 2 \begin{bmatrix} 1 & 0 \\ 0 & 0 \end{bmatrix} + \frac{-4}{2} \begin{bmatrix} 0 & 1 \\ -1 & 0 \end{bmatrix} + (-3) \begin{bmatrix} 0 & 0 \\ 0 & 1 \end{bmatrix}$$

$$= \begin{bmatrix} 2 & -2 \\ 2 & -3 \end{bmatrix}.$$

39. We apply the Gram-Schmidt process to the basis $\{\mathbf{u}_1, \mathbf{u}_2, \mathbf{u}_3\} = \{1, x, x^2\}$ to obtain an orthogonal basis for W. Let $\mathbf{v}_1 = \mathbf{u}_1 = 1$,

$$\mathbf{v}_2 = \mathbf{u}_2 - \frac{\langle \mathbf{u}_2, \mathbf{v}_1 \rangle}{\|\mathbf{v}_1\|^2} \mathbf{v}_1$$

$$= x - \frac{\int_0^1 x\,dx}{\int_0^1 1^2\,dx} 1 = x - \frac{1}{2},$$

and

$$\mathbf{v}_3 = \mathbf{u}_3 - \frac{\langle \mathbf{u}_3, \mathbf{v}_1 \rangle}{\|\mathbf{v}_1\|^2} \mathbf{v}_1 - \frac{\langle \mathbf{u}_3, \mathbf{v}_2 \rangle}{\|\mathbf{v}_2\|^2} \mathbf{v}_2$$

$$= x^2 - \frac{\int_0^1 x^2\,dx}{\int_0^1 1^2\,dx} 1 - \frac{\int_0^1 x^2 (x - \frac{1}{2})\,dx}{\int_0^1 (x - \frac{1}{2})^2} \left(x - \frac{1}{2} \right)$$

$$= x^2 - \frac{1}{3} - \left(x - \frac{1}{2} \right) = x^2 - x + \frac{1}{6}.$$

Next, we divide each $\mathbf{v}_i$ by its norm to obtain the orthonormal basis

$$\{\mathbf{w}_1, \mathbf{w}_2, \mathbf{w}_3\} = \left\{ 1, \sqrt{3}(2x - 1), \sqrt{5}(6x^2 - 6x + 1) \right\}$$

for W.

41. We use the orthonormal basis $\{\mathbf{w}_1, \mathbf{w}_2, \mathbf{w}_3\}$ derived in Exercise 39 to obtain the desired orthogonal projection. Let $\mathbf{w}$ denote the function $\mathbf{w}(x) = \sqrt{x}$. Then

$$\mathbf{w} = \langle \mathbf{w}, \mathbf{w}_1 \rangle \, \mathbf{w}_1 + \langle \mathbf{w}, \mathbf{w}_2 \rangle \, \mathbf{w}_2 + \langle \mathbf{w}, \mathbf{w}_3 \rangle \, \mathbf{w}_3$$

$$= \left(\int_0^1 1\sqrt{x} \, dx \right) 1 + \left(\int_0^1 \sqrt{3}(2x - 1)\sqrt{x} \, dx \right) \sqrt{3}(2x - 1)$$

$$+ \left(\int_0^1 \sqrt{5}(6x^2 - 6x + 1)\sqrt{x} \, dx \right) \sqrt{5}(6x^2 - 6x + 1)$$

$$= \frac{2}{3} + \frac{2}{5}(2x - 1) + \frac{-2}{21}(6x^2 - 6x + 1)$$

$$= \frac{6}{35} + \frac{48}{35}x - \frac{4}{7}x^2.$$